This Year in School Science 1989

National Forum for School Science.

Scientific Literacy

Papers from the 1989
AAAS Forum for School Science

Edited by

Audrey B. Champagne
Barbara E. Lovitts
and
Betty J. Calinger

American Association for the Advancement of Science
Washington, D.C.

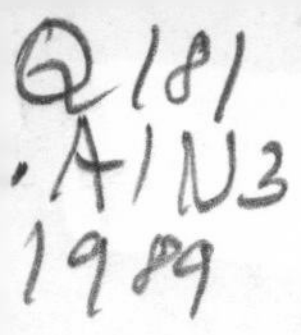

The AAAS Board of Directors, in accordance with Association policy, has approved the publication of this work as a contribution to the understanding of an important area. Any interpretations and conclusions are those of the authors and do not necessarily represent views of the Board or the Council of the Association.

ISBN 0–87168–359–8
ISSN 0888–4358
AAAS Publication 89–30S

Printed in the United States of America

This Year in School Science 1989

Scientific Literacy

About the Forum

The American Association for the Advancement of Science (AAAS) Forum for School Science is part of the Directorate for Education and Human Resources Programs of AAAS. Each year, the Forum collects and analyzes information about a particular issue in science education and sponsors discussions on that topic. Volumes in the series, *This Year in School Science,* extend the information and discussion to those who are unable to attend the Forum meetings and preserve the Forum's deliberations for further examination. The Forum operates on a three-year cycle (focusing in turn on the topics of Science Teaching, the Science Curriculum, and Science Learning), in order to monitor progress in policy and practice over time.

The Forum project is funded in part by the Carnegie Corporation of New York with additional support from AAAS.

Program staff:

Audrey B. Champagne, *Program Director*
Barbara E. Lovitts, *Program Associate*
Betty J. Calinger, *Program Administrative Associate*
Gary E. Hammond, *Program Secretary*

Advisory Board

Contents

Preface

This volume marks both the fifth year of the American Association for the Advancement of Science's (AAAS's) Forum for School Science and nearly a decade of sustained national attention to the quality of science education. The 1983 report by the National Science Board Commission on Precollege Education in Mathematics, Science, and Technology, *Educating Americans for the 21st Century*, first formally expressed the science and education communities' concern about the quality of school science instruction. Today, the private sector is expressing similar concerns. Meanwhile, efforts to document the nature and extent of the problem are evolving into attempts to change school science curricula in order to improve the scientific literacy of high school graduates.

According to the critics who cite science achievement data collected here and abroad, we are a nation of scientific illiterates. Even our brightest and best are not competitive with the science students of other industrialized nations. These include our major competitors, such as Japan and the European community. Consequently, the call for improved science education centers on the nation's need to maintain its competitive edge.

Scientific literacy is a familiar phrase in the rhetoric of reform. Unfortunately for education, the phrase, as sometimes used, is essentially an empty one. The public expects its schools to engender scientific literacy, but does not define it in a way that will help educators to achieve it. The premise of Forum 89 is that, without agreement on definition, our efforts to achieve scientific literacy are doomed to failure.

The goal of Forum 89 is to contribute to national efforts to define scientific literacy. To this end, the chapters in this volume seek to capture elements of the diverse ideas about the characteristics of scientific literacy, illuminate some reasons for the variety, and consider the implications of that diversity on changes in school science curricula. The authors – a philosopher, an economist, scientists who are educators, and educators who are scientists – reflect the multidisciplinary nature of the definition problem. Our hope is that their various perspectives will encourage reflection and stimulate debate that will produce clearer mutual understanding of what it means to be scientifically literate.

Support for the publication of this volume and the Forum meeting comes from the Carnegie Corporation of New York and the AAAS membership. We thank both Carnegie and the AAAS for their contributions to this and other projects undertaken to strengthen and broaden participation in science education.

Audrey B. Champagne
Barbara E. Lovitts
Betty J. Calinger

Contributors

John Bishop
Associate Professor
School of Industrial and Labor Relations
Cornell University
Ithaca, NY

Nancy W. Brickhouse
Assistant Professor
School of Education
University of Delaware
Newark, DE

Audrey B. Champagne
Senior Program Director
Directorate for Education and Human Resources Programs
American Association for the Advancement of Science
Washington, DC

Angelo Collins
Director, Teacher Assessment Project and
Assistant Professor
School of Education
Stanford University
Stanford, CA

Diane Ebert-May
Assistant Provost for Special Sessions
University of Delaware
Newark, DE

Gerald Fourez
Professor and Chairman
Department of Science, Philosophy, and Society
University of Namur
Namur, Belgium

Paul DeHart Hurd
Professor Emeritus of Science Education
Stanford University
Stanford, CA

Barbara E. Lovitts
Program Associate
Directorate for Education and Human Resources Programs
American Association for the Advancement of Science
Washington, DC

Morris H. Shamos
Professor Emeritus of Physics
New York University
New York, NY

Betty A. Wier
Assistant Professor
School of Education
University of Delaware
Newark, DE

1

Scientific Literacy: A Concept in Search of Definition

Audrey B. Champagne and Barbara E. Lovitts

Scientific literacy is the catch phrase of the educational discourse of the 1980s. National reports from the educational, governmental, and private sectors call upon the nation's schools to improve school science education and, as one consequence, stem the decline of the American economy. Although the reports agree that a scientifically literate citizenry is important to the nation, all fail to describe (in a way that enables measurement) what it means to be scientifically literate. Based on the premise that that which is not defined adequately cannot be pursued effectively, the authors of the chapters in this volume consider different perspectives of scientific literacy. Their goal is to help schools meet national expectations for a scientifically literate populace. For this reason, we begin this volume with a look at the conceptual barriers to achieving consensus on what constitutes scientific literacy and at the implications for school science curricula of our inability to describe clearly what we expect from the schools.

Our analysis of the conceptual barriers is based on the framework depicted in Figure 1. This framework illustrates three broad categories into which the descriptive elements of scientific literacy can be placed. It also shows the influence of ideology on the nature of the descriptions. Responses to the question: "What does it mean to be scientifically literate?" typically include descriptions of the behaviors of scientifically literate persons in a variety of contexts, the mental state of a scientifically literate person—knowledge, skills, and dispositions,—and references to educational experiences that are assumed to produce a scientifically literate person. In this framework, ideology influences

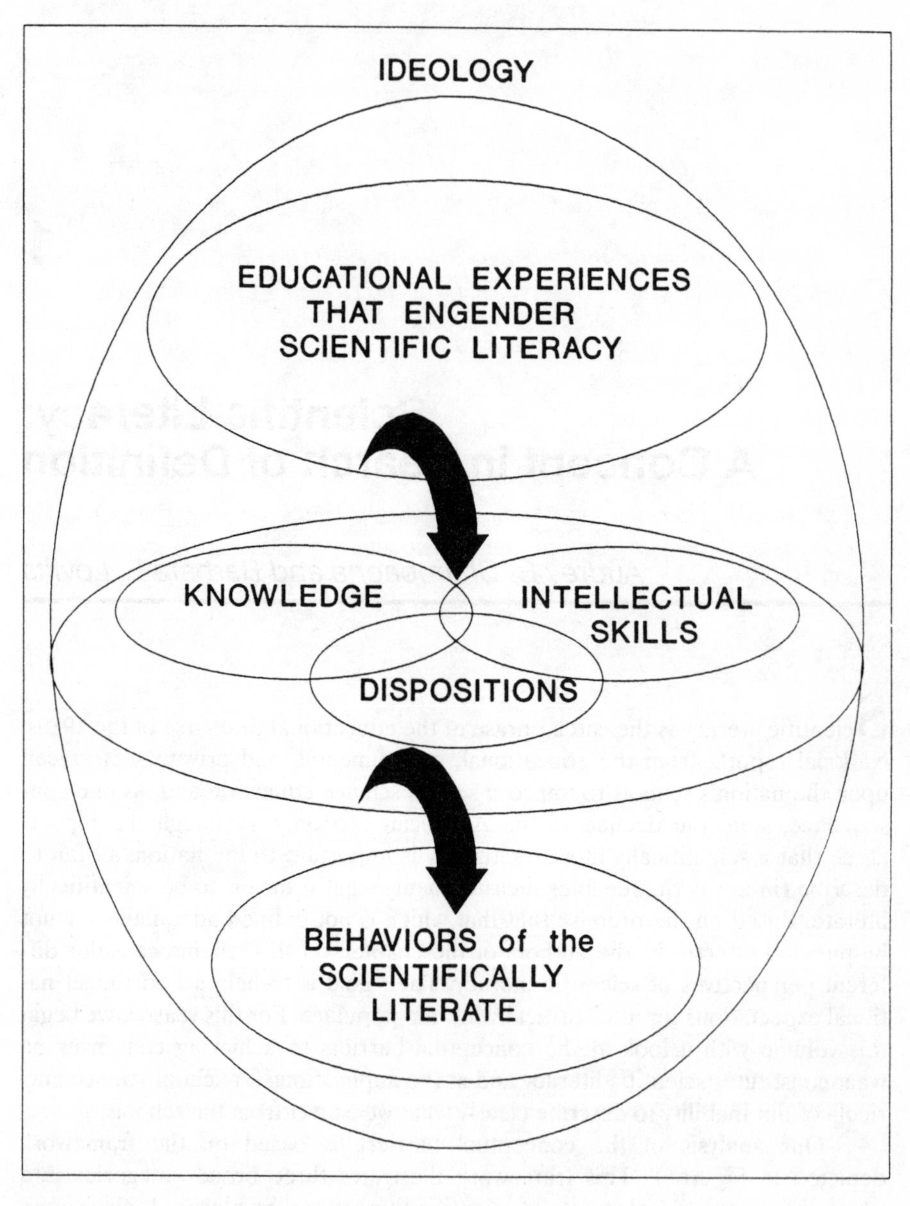

Figure 1
Conceptual Framework for Scientific Literacy

the selection of the category–conditions of learning, mental state, or behaviors–on which a definition will be based. The way in which selected aspects within a category are represented is also influenced by ideology.

Barriers to Consensus

Relationship of Characteristics and Experiences

In their paper, *Visceral Priorities: Roots of Confusion in Liberal Education,* Ahlgren and Boyer (1981) observe that "in discussions of the ideal education, educators often seem to be not so much disagreeing as to be talking past one another–each belaboring issues that the others consider unimportant" (p. 174). Ahlgren and Boyer illuminate important distinctions in the nature and purposes of education that often go unrecognized and result in the perception that serious and seemingly irreconcilable differences exist with respect to the nature and purposes of education. These differences, however, are only of emphasis on either nature or purpose. The visceral confusion among educational purposes, course content, instructional methods, and alleged changes in students are typical of educational discourse and represent a formidable barrier to achieving consensus on descriptions of scientific literacy. An illustration of the ways in which the intermingling of purposes, means, and outcomes obscures the issues in reaching consensus on scientific literacy is presented in Chapter 7 by Nancy Brickhouse, Diane Ebert-May, and Betty Wier. In that chapter, scientists, in particular, are shown responding to questions about outcomes (characterizations of the scientifically literate person) with prescriptions for means (the conditions under which they are supposedly achieved). We concur with Ahlgren and Boyer that, if distinctions are made among purposes, means, and outcomes, then "arguments may be no more fruitful or less heated, but they may be less frustrating" (p. 174).

Diversity and Complexity of Behaviors, Knowledge, Skills, and Dispositions

The diversity and complexity of the behaviors, knowledge, intellectual skills, and dispositions considered to characterize the scientifically literate person also serve as barriers to achieving consensus. To some, being scientifically literate means the appropriate application of scientific knowledge and reasoning skills to solving problems and making decisions in one's personal, civic, and professional affairs. For others, being scientifically literate means the ability and inclination to continue learning about science lifelong.

For still others, scientific literacy is often equated with being knowledgeable about science and having certain intellectual skills, whether they are used or not. Scientific knowledge has several components, however, and disagreements frequently take place over which components are most important. One component comprises scientific concepts, principles, laws, and theories. A second component relates to the nature of scientific inquiry and its philosophical underpinnings. A third component includes the context of science, its historical development, and the interactions of science, society, and technology.

Judgments about which of these aspects of scientific knowledge it is necessary for everyone to know are based in part on perceptions of the contributions that knowledge can make to the ability to participate in contemporary American life. These judgments will change over time because the knowledge required to function in society changes. Varying judgments about the relative importance of knowing the products of scientific inquiry, the processes of scientific inquiry, and the historical, cultural, and philosophical contexts of science will produce very different descriptions of the characteristics of scientifically literate persons.

Equally extensive and diverse are the intellectual skills that contribute to scientific literacy. Such intellectual skills include those associated with the practice of scientific inquiry and the solution of both academic and actual scientific problems as well as those that enable an individual to learn more about science.

The disposition to value and the propensity to apply scientific knowledge and intellectual skills are also elements of scientific literacy. However, considerable disagreement exists about the relative importance of these elements. The issue is whether having a knowledge of science makes one scientifically literate in the absence of appreciating science or applying scientific knowledge outside the classroom.

Ideology and the Relative Importance of Aspects of Scientific Literacy

Ideology refers to the images of the world that influence individuals' social values and judgments. The images structure perception, determining those aspects of the environment to which a person will attend and with which he or she will be concerned. Ideology also legitimates patterns of action for the individual either by explicitly assigning value – for example, "Democracy is a good social system" – or by presenting them as if they are obvious – for instance, "Boys do not cry" (Fourez, 1988 p. 269). These images result in an individual's promotion of particular values or world views, often without explicit reference to the underlying ideology. Ideologies influence descriptions of scientific literacy. Descriptions of how the scientifically literate person should think and act are conceptualized within the context of an image (generally tacit) of the ideal citizen functioning in an ideal society. The fact that the assumptions about the context are neither stated nor negotiable is a source of frustration in efforts to reconcile differences in descriptions of scientific literacy.

Quite different definitions of scientific literacy arise because individuals assign different values to knowledge. Whether scientific knowledge is esteemed primarily for the pleasure it provides the individual or for its utility will be determined by ideology and results in differing definitions of scientific literacy. Similarly, images of the relationship between the individual and society influence conceptions of the purposes of scientific literacy. For instance, conflicting definitions of scientific literacy can be attributed to different views on whether scientific literacy is primarily for the benefit of the individual or for the benefit of society. The contemporary rationale for scientific literacy is often embedded in the national need for a more scientifically competent work force. Consequently, in the contemporary view, the contribution of scientific literacy to the pleasure and power of the individual is subordinate to its importance to the nation.

The significance of intellectual skills in scientific inquiry is of particular interest with regard to ideology. If the image of the ideal society is one in which the existing social and political order is to be maintained, little value will be assigned to inquiry skills that intentionally lead to the questioning of authority and that seek to reexamine society's understanding of the natural world. Similarly, if the image of the ideal society includes science as the route to the solution of social problems, then greater value will be placed on intellectual skills that stress objectivity and rationality than on those skills necessary for social and political negotiation.

Relationship of Knowledge and Intellectual Skills to Behavior

The effect of this barrier is exemplified in the responses by scientists, educators, teachers, students, and science policy analysts to an informal national survey conducted by the American Association for the Advancement of Science (AAAS). The questionnaire asked respondents to rate the importance of 15 capabilities and attitudes to scientifically literate high school graduates on a seven-point scale ranging from 1 (not necessary) to 7 (essential). The 5 capabilities consistently rated highest are *read-and-discuss abilities* that contribute to informed personal and civic decisionmaking as well as competent performance in the workplace (Figure 2). The capabilities consistently rated lowest are *define, describe, and design* abilities which are commonly associated with answering examination questions in a school setting. These capabilities

---

Read and understand articles on science in the newspaper.

Read and interpret graphs displaying scientific information.

Engage in a scientifically informed discussion of a contemporary issue, e.g., should a child with AIDS be allowed to attend public school.

Apply scientific information in personal decisionmaking, e.g., ozone depletion and the use of aerosols.

Locate valid scientific information when needed.

Figure 2
Highest-Rated Capabilities Contributing to Scientific Literacy

Provide a scientific explanation for a natural process, e.g., photosynthesis, digestion, combustion.

Assess the methodology of an experiment.

Define basic scientific terms, e.g., DNA, molecule, electricity.

Design an experiment that is a valid test of a hypothesis.

Describe natural phenomena, e.g., the phases of the moon.

Figure 3
Lowest-Rated Capabilities Contributing to Scientific Literacy

require knowing about science, its knowledge products, and the methods that produce them (Figure 3).

The *academic* capabilities, however, would seem to be necessary components of the capabilities rated essential to being scientifically literate. (Obviously, it is not possible to understand a newspaper article about scientific issues – the potential benefits or dangers of gene splicing, for instance – without knowing the meaning of the scientific terms used in the article – gene, DNA, and molecule, for example, or without possessing some knowledge of biological systems.) What is clear from the ratings is that the "academic abilities" are not widely considered to be ends in themselves, however relevant they may be as means to interpretation and application.

Describing the relationship between academic knowledge about science and the competencies of adults is essential if we are to be successful in reaching our national goal of a scientifically literate populace. The description requires the identification of information about science and of the intellectual skills that enable the scientifically literate adult to exhibit the desired competencies.

The academic knowledge and skills required for adult competence can be inferred from detailed descriptions of the scientifically literate adult. This process can be illustrated by using an example of an adult capability from the AAAS Scientific Literacy Questionnaire: Read and understand articles on science in a newspaper. What does it mean to read a newspaper article with understanding? Understanding implies the ability to assess the reliability of the source of the information. An ideal citizen would know that *The New York Times* is a more reliable source of information about science than *The National Enquirer. The Times* is more reliable in two respects – first *The Times* screens originating sources of information and second *The Times* reports accurately. Understanding also implies that the ideal citizen is alert to potential bias, especially where reports on scientific work have implications for social or political action. This competence requires knowing about the political and social views of a newspaper's publisher as well as being alert to the social and political implications of the results of particular lines of research. Understanding a newspaper article also implies the ability to detect factual errors, inappropriate experimental design, unwarranted inferences from data, and poorly structured arguments.

This more detailed description of competence clarifies the relationship of knowledge about science and science-related reasoning skills to the ability to read and understand articles about science. Given the above definition of understanding, knowledge about science, ranging from familiarity with the meaning of basic terms to recognition of the relationships between science and society, is required in order to be considered scientifically literate. In addition, this level of understanding requires highly sophisticated reasoning skills.

Do we as a nation expect that all citizens will develop the level of scientific knowledge and reasoning skills necessary to understand newspaper articles to the extent described? If not, how do we begin to define alternative, more modest levels?

Levels of Scientific Literacy

With regard to the matter of assessing the reliability and the social and political views of newspapers, do we expect the ideal citizen to make independent judgments or simply to rely on the assessment of others, in which case the measure of scientific literacy will be only whether the person believes recommended newspapers exclusively. The schools' responsibility then would be to provide a list of newspapers and magazines with good reputations for science reporting and to encourage students to read them. Which do we expect of our ideal citizen – the capacity for independent judgment of the scientific validity of reports about science or the habit of reading recommended periodicals? Or, in fact, do we expect the ideal citizen to be capable of making certain kinds of judgments at a lower level of sophistication. If so, which judgments should the schools help to develop and at what level of sophistication?

What implications do choices among levels of competence have for the scientific knowledge and intellectual skills that schools are expected to develop? If, for instance, a person reads an article and recognizes that the issue is an important one, but that his or her knowledge of the scientific content is insufficient to understand the article, is it our expectation that the person will know where to go in order to become knowledgeable about the topic and will have the basic background knowledge and learning skills to take advantage of the source, or is it sufficient to recognize one's adequacy or inadequacy to judge? In this regard, we have often heard experts say that they are not concerned that young people learn a lot of information about science because they can always learn the relevant scientific content when they need it. However, the ability to recognize that one's information is insufficient and then to seek enlightenment requires having some scientific information. The question becomes, what is that information?

Another perspective on the appropriate level of scientific understanding for an ideal citizen relates both to knowledge about the natural world and the scientific processes that are applied to gain that knowledge. Knowledge about the natural world can be phenomenological, experimental, or theoretical. At the phenomenological level, knowledge is obtained by observation, is qualitative in nature, and is communicated as descriptions. At the experimental level, knowledge involves measurement and results in correlational relationships. At the theoretical level, knowledge is expressed in terms of causal relationships. The issue relevant to scientific literacy is, what should be the level of our ideal citizen's knowledge about science? Is the level different for topics at the frontier of science – chaos, for instance – than for basic topics – the particulate nature of matter or equilibrium processes, for example? Or is it more important to be aware that understanding a topic or concept requires knowledge about the natural phenomena related to it and the experiments used to collect data about it, in addition to knowing the theory? What level of scientific understanding empowers the individual and allows him or her to gain pleasure from understanding the natural world? What level of scientific understanding allows one to engage competently in personal, civic, and social activities and to contribute to the economy?

Perspectives on Scientific Literacy

As a consensus definition of scientific literacy is hammered out, several questions must be kept in mind: What is the nature of the society in which the person will function? What purpose will scientific literacy serve in that society? Who should be scientifically literate and to what degree? What knowledge is needed in order to be scientifically literate? Why is that specific knowledge necessary? What individuals or groups are responsible for defining scientific literacy and what is their interest in the outcomes? And, finally, how can scientific literacy be assessed? These are some of the questions addressed in the chapters of this volume.

The Ideological Field in the 20th-Century American Context

In Chapter 2, Paul Hurd describes the ideological field in which late 20th-century Americans are trying to define scientific literacy. Hurd contends that the definition of scientific literacy that is currently guiding educational practice is out of step with the realities of living, learning, and working in modern America and that existing science curricula are socially, culturally, and cognitively outdated. Society has shifted from being industrially-based to being information/service-based, concomitant with a trend throughout this century to divorce work from schooling. The new definition of scientific literacy called for is premised on the economic needs of the nation – more specifically, on the needs of business and industry and on new conceptions of the workplace and the worker. Scientific literacy is essential because science and technology are both the engines of economic change and the raw materials of international commerce and because policy research indicates that the proportion of jobs that require technical training is increasing. Therefore, the quality of human capital must be improved in order for individuals to be more productive in their personal lives and in the workplace so that America can compete more effectively in the global economy.

Hurd argues that, to be scientifically literate in this context, students must be equipped with the cognitive and noncognitive skills relevant to their occupational futures and possess the temperament for acquiring such knowledge and skills. These skills include the ability to solve problems encountered in the workplace and in the conduct of personal and civic life, communicate effectively, work in groups, and work independently. Hurd concludes that achieving consensus on the essential components of scientific literacy requires the cooperative efforts of those concerned with and informed about science education—people who are informed about the natural, cognitive, and social sciences, technology, education, and business.

Who Needs to Be Scientifically Literate?

At Forum 86, Daniel Koretz, a Congressional budget analyst, stated that he knew of no convincing evidence linking science education and economic productivity (Champagne & Hornig, 1987, p. 3). Similarly, Rosalie Cohen (Champagne & Hornig, 1987, p. 3; Cohen 1987) claimed that, contrary to predictions of a technology- and information-based economy, most workers will be in the service sector; therefore, not only will they not need the knowledge and intellectual habits valued by scientists and promoted in science courses but also they actually will find them counterproductive. These claims are challenged in Chapter 3, in which John Bishop presents empirical evidence linking job performance and scientific competence and calls for restructuring school-based and labor-market rewards for studying science—a strategy that he asserts will encourage young people to become scientifically literate.

At first glance, Bishop argues, achievement in science courses appears to have no special correlation with wage rates, earnings, or unemployment of young men and women when other competencies are held constant. Many employers report that their firms never use the knowledge that an individual gained in high school science courses. However, research using the Armed Services Vocational Aptitude Battery indicates that scientific knowledge has modest correlations with worker performance on military jobs that are similar to blue-collar and technical jobs in the private sector. Because achievement in science courses is not generally considered in hiring entry-level workers, employers do not realize how scientific knowledge and skills contribute to worker productivity; therefore, they cannot learn from experience which scientific skills are helpful on the job.

Bishop argues that students are apathetic about learning science because our society fails to recognize and reward those who have studied it. High school graduates' knowledge of science is not communicated adequately to colleges or employers and, consequently, does not influence college admission or hiring decisions significantly. The key to motivating students to take science courses is to recognize and reward their learning achievements. College admission criteria need to be revised to place more emphasis on learning science. Employers need to know who is competent in science and schools need to communicate their students' achievement in science to prospective employers. In addition, schools need to revise their science curricula. They need to teach students concepts and skills they will use after high school and to present science as a useful means of finding solutions to practical problems. Schools, too, need to establish a better system of rewards for studying science. Only when all students, not just those planning to pursue science-related careers, perceive personal benefit from learning science will they be motivated to become scientifically literate.

Conceptions of Scientific Literacy in National Reports

The ideology implicit in conceptions of scientific literacy has important implications for the design of science curricula and for the perceptions of science and perceptions of the relationship between science and society that students will develop. These perceptions, in turn, will influence the ways students will think about, and act with respect to, science throughout their lives. In Chapter 4, Gerard Fourez examines four national reports which will influence future science curricula—*Science for All Americans* (American Association for the Advancement of Science, 1989), *Science Objectives: 1990 Assessment* (National Assessment of Educational Progress, 1989), *Science and Technology Education for the Elementary Years* (National Center for Improving Science Education, 1989), and *Science for Ages 5 to 16* (Department of Education and Science and the Welsh Office, 1988). For each document, Fourez infers the ideological positions on scientific literacy, discusses the bases for choices contained in the different positions, and considers the implications of these choices for the individual and for society.

Each document presents a different view of what it means to be scientifically literate. The position on scientific literacy in one document derives from students' questions about their environment and their lives, whereas in another the notion of scientific literacy comes from societal expectations. All views of scientific literacy are influenced also by the authors' perceptions of the nature of science. Some portray science as *empiricist*—an attitude that minimizes the human and historical contexts of scientific development—and *technocratic*—an assumption that science produces solutions to societal problems without human or political negotiation. In other reports, science is portrayed as a human process that involves people seeking to adapt to their environment.

Similarly, definitions of scientific literacy are influenced by conceptions of the relationship between science and society. One report, Fourez finds, places scientific literacy in the same ideological field as does Hurd. It emphasizes the importance of scientific literacy for competition with other nations and for improving the state of the work force in the United State. Another report posits a world in which scientific literacy ameliorates the social and economic inequalities among groups typically underrepresented in science – women and racial and ethnic minorities. A third report purportedly reflects the dominant beliefs of America's individualistic and technocratic culture, while the fourth highlights the projected demographic changes in American society and underscores the importance of educating students in such a way that they will be willing to make the necessary changes in their lives to solve societal and environmental problems. Clearly, the varying conceptions of scientific literacy, science, and the relationship between science and society contained in each document will not be neutral with respect to their influence on curricula or with regard to the products of the curricula, that is, students' understanding of science and their applications of science in daily life.

Defining Scientific Literacy at the Elementary School Level and Assessing Curricula Designed to Achieve It

New emphasis is being placed on elementary school science education. In recent years, educators have become convinced that the elementary school years are the time when the greatest impact can be made in science education because children's curiosity is at its peak and they are most receptive to new ideas. Morris Shamos (Chapter 5) and Angelo Collins (Chapter 6) independently assess National Science Foundation-sponsored elementary school science curricula. Their objectives in evaluating the curricula are different: Shamos, to assess the adequacy of the new curricula to achieve *true scientific literacy* and Collins, to develop a strategy or process to assist teachers and science supervisors in selecting elementary school science curricula. But both evaluate the curricula in light of their idiosyncratic views of what it means to be scientifically literate and how best this literacy can be attained.

Shamos places the new curricula in the context of past efforts to reform elementary school science curricula and bases his assessment of the new curricula on his conception of true scientific literacy. For Shamos, the person who has achieved true scientific literacy understands the processes of science and the role of scientific methods in coming to understand the natural world. The scientifically literate person also knows something about the scientific enterprise – the major conceptual schemes of science, how they were arrived at, and why they are widely accepted. Given this definition, Shamos believes that the new National Science Foundation-sponsored elementary school science curricula will be no more effective in developing true scientific literacy than the curricula developed during the reform movement of the 1960s and 1970s. In fact, he believes the new curricula are less likely to succeed because they do not emphasize the processes and foundations of science, they do not stress how laws and theories may be used to account for the observed facts of nature, and they do not develop a comprehensive view of the scientific enterprise. However, Shamos acknowledges that true scientific literacy, as he defines it, is an unreasonable expectation as an immediate goal for elementary school science (and for most of the public), but he concedes that the curricula will foster some degree of cultural and functional scientific literacy, that is, students will develop a command of a scientific vocabulary and be able to converse, read, and write credibly about science and know some of the simple everyday facts of science.

By contrast, Collins takes the perspective of an elementary school teacher or district science supervisor who must choose among several curricula, all claiming to engender scientific literacy. For Collins, scientific literacy is a desired level of depth and breadth of scientific understanding appropriate to the interests and needs of the students being taught, set within the context of the needs of the community in which the student lives. This conception of scientific literacy is based on a multifaceted conception of the nature of science. Science, as conceived by Collins, is constructed by humans to describe, understand, explain, predict, and control natural phenomena. It contains three components: a structural component – events, facts, concepts, relationships, theories, and models; a procedural component – a repertoire of skills necessary to manipulate the elements of scientific knowledge and build conceptual structures, and a human component – the relationship of the sciences to each other and to other disciplines, as well as the relations and connections between science and daily life. Achieving scientific literacy, Collins contends, requires using curricula that place a balanced emphasis on all three components.

Conceptions of Scientific Literacy: Public Consensus or Dissent?

Achieving a consensus definition of scientific literacy begins with ferreting out the ideas, conceptions, and images that various concerned sectors of the public have of scientific literacy and the scientifically literate citizen. Viewpoints need to be compared both within and between sectors. Brickhouse, Ebert-May, and Wier (Chapter 7) present an analysis of transcripts from a series of roundtable discussions, on the nature of scientific literacy, attended by four groups involved in science education – teachers, school administrators, pre-service teachers (college students), and industrial chemists. The points of consensus and dissent that they report are as illuminating as they are disconcerting.

Each group as a whole tended to speak with one voice, but the voices of the groups were identifiably different. The groups agreed at a very general level that scientifically literate persons should be aware of and possess skills necessary to deal with the science-related, social and personal issues they encounter in everyday life. However, they failed to agree not only on the details of the ways in which scientifically literate people should behave but also, more significantly, they disagreed on the reasons for those competencies that they identified as necessary. In sum, the analysis reveals a great discontinuity between that which the four groups believe is the essence of scientific literacy and the nature of science taught in the schools today.

Conclusion

Although the concerned public holds seemingly divergent views on the purposes of scientific literacy, the content and the methods of science courses needed to achieve it, the characterization of the scientifically literate person, and even who needs to be scientifically literate and to what degree, one thing is certain: Scientific literacy is essential to everyday life. But, in practice, scientific literacy is currently conceptualized as academic as well as intellectual knowledge and skills. This interpretation is consistent with that of the contributors to this volume, the findings of the round-table discussions presented in Chapter 7, and the AAAS Scientific Literacy Questionnaire. Overall, the contributors, discussants, and respondents are most concerned with those aspects of scientific literacy that impinge on everyday life and least concerned with those aspects that may be considered "school skills"—the knowledge and skills on which students are commonly assessed. It appears that there is a major gap between what we as a society desire in the scientifically literate person and the way in which we educate the populace in and about science. This gap arises in part from the different conceptualizations the public has about the nature of late 20th- and early 21st-century society and about what the scientifically literate person needs to know and be able to do in order to function in that society and lead a rewarding and fulfilling personal, professional, and civic life.

References

Ahlgren, A., & Boyer, C. M. (1981). Visceral priorities: Roots of confusion in liberal education. *Journal of Higher Education, 52*(2), 173–181.

American Association for the Advancement of Science. (1989). *Science for all Americans: A Project 2061 report on literacy goals in science, mathematics and technology*. Washington, DC: Author.

Champagne, A. B., & Hornig, L. E. (1987). Critical questions and tentative answers for the school science curriculum. In A.B. Champagne & L. E. Hornig (Eds.), *This year in school science 1986: The science curriculum* (pp. 1–12). Washington, DC: American Association for the Advancement of Science.

Cohen, R. A. (1987). A match or not a match: A study of intermediate science teaching materials. In A. B. Champagne & L. E. Hornig (Eds.), *This year in school science 1986: The science curriculum* (pp. 35–60). Washington, DC: American Association for the Advancement of Science.

Department of Education and Science and the Welsh Office. (1988). *Science for ages 5 to 16.* London: Author.

Fourez, G. (1988). Ideologies and science teaching. *Bulletin of Science, Technology, and Society. 3*(8), 269–277.

National Assessment of Educational Progress. (1989). *Science objectives: 1990 assessment.* Princeton, NJ: Educational Testing Service.

National Center for Improving Science Education. (1989). *Science and technology education for the elementary years: Frameworks for curriculum and instruction.* Washington, DC: Author.

2

Science Education and the Nation's Economy

Paul DeHart Hurd

Over the past decade, the public, business, and the federal government have recognized that complex changes are taking place in the nation. These changes – the accelerated growth of new knowledge in the sciences, new technologies, a new global economic order, new conceptions of work, new uses of leisure time, and changing demographics – all have implications for science education.

Since 1983, over 300 reports on the condition of education in the United States have been issued. Inherent in the majority of these reports is a concern that, due to changes in society, education has lost contact with society, the workplace, and the economy. These reports stress that it is time to reformulate the goals of science education to enable individuals to be more adaptable to change and the nation to be competitive and productive in the evolving world economy. What is sought is a new social contract between schools and society – a contract that recognizes the realities of living, learning, and working in a changing society.

As individuals and as a nation, we have entered an era of rapid and unpredictable change, more turbulent and with an orientation different from any we have experienced in the past. Consequently, understanding the nature of change, being able to cope with it, and having the ability to adapt successfully to constant change have become educational imperatives. It is within this context that I perceive a rationale for scientific literacy as an educational goal – one that incorporates a framework of actions that will increase a person's potential for living, learning, and working in a changing society. However, we are still far from achieving a coordinated national policy to realize this goal.

Unproductive Steps

Most of the efforts to improve science education over the past six years have resulted only in intensifying the pattern of conditions that gave rise to the demands for reform. Many of the changes have been reactions to the symptoms rather than reconceptualizations for a new era. Examples of the structural changes implemented already include: lengthening the school day and year, requiring more science courses, intensifying course rigor, increasing student testing and school assessments, and raising graduation requirements; but, to what ends? Curricular changes have taken the form of slogan-driven appeals for "quality," "excellence," "scientific literacy," and "critical thinking"—terms that have been largely undefined.

The slogans have been accompanied by a flow of "purple" rhetoric: unanalyzed assumptions, statistics showing that we are losing ground in the achievement of traditional goals, and assertions that we should judge our scholastic results by comparisons with foreign countries. However, there is no discussion of the fact that these foreign countries are reforming their science curricula to accommodate changing economies. Nor have educational policy analyses paid sufficient attention to fluctuating social and economic conditions. Research on science education policy has taken scant account of social and economic vicissitudes, the products of modern science and technology, developments in the cognitive sciences, cultural shifts, and the changing workplace. The little policy research that has been done, lacks systematic analysis, synthesis, or reflection on present conditions in light of the nation's economic and educational history.

The Nature of Policy Analysis

In general terms, policy analysis is concerned with the development of guidelines for resolving problems and issues and with recommendations for action. Determining policies for science education requires looking beyond the quantitative measures of achievement tests, enrollment figures, course offerings, and the supply of teachers and their qualifications. It must include analyses of the larger system as it is and as it is envisioned. Policy studies in education must focus on the panorama of everyday human activities in relationship to the broader society and the forces that are fostering changes in the ways that people live, learn, and work.

Neither individuals nor groups can be expected to perceive issues in the same way or to propose identical solutions. Comparisons of the many national reports on the condition of education in the United States attest to the variety of viewpoints. Even so, it is important that all those with concerns have an opportunity to present them. But now, after six years of professional, political, and public pronouncements, we must pool our insights in order to identify and agree upon a unifying focus; otherwise we cannot expect to achieve lasting and effective reforms in science education.

The themes that emerge from an analysis of science education policy studies provide the basis for a conceptual framework that can guide the immediate future of science education. Not all of the themes are fixed; rather, they are manifestations of today's society and culture. But in placing science education in the configuration of existing culture and making plans based on our best prediction of the future, history must not be forgotten. New policies do not necessarily require total discard of traditional beliefs, habits, and conditions. Rather, history allows us to (a) identify past conditions that led to public appeal for new perspectives, (b) provide logic for change, and (c) identify what of the past has proven worthy of retention for the future. History also reminds us that certain educational issues are perennial. One such issue is whether the purpose of schooling is to change society or to transmit norms and values of society in their present form.

Boundaries of Policy Analysis

Often, schools are expected to function both as agents of social change and as keepers of the existing social and economic order. Therefore, policy for the reform of science education must be formulated in recognition of the fact that there are always social and cultural elements that schools can do little to alter. The continuous berating of education in the United States is as much an expression of social frustration about conditions that the schools cannot change as it is an expression of dissatisfaction with the schools and schooling. Even in the formulation of policy to modify traditional patterns of science education in line with various cultural shifts, economic changes, and advances in science and technology, only a few of these factors can be addressed simply by reconceptualizing science education. As we look to the future, the number and importance of forces that will have effects on education are formidable.

According to Oxford Analytica (1986) – a group of scholars at Oxford University who specialize in assessing the implications of national and international developments in business – the social, economic, political, fiscal, and psychological trends likely to shape American society for the next 10 years and beyond include:

- The rise of neo-conservatism in politics
- Our high risk and high-stress society
- New patterns of family life
- The increasing rate of downward mobility

- The diffusion of political power
- Trends toward class stratification
- Resistance of minority groups to assimilate into the national life
- A likely low-growth economy over the next decade
- The concentration of jobs in the service and high technology sectors
- Changes in American lifestyles and habits resulting from modern technology
- The growing number of women in the labor market
- Cooperative socialism
- Changing ethnic demography
- The extent of federal support for research and development in science and technology.

While these factors are not to be ignored in restructuring science education, their implications are subject to debate and remain largely unresolved.

Knowledge and the Economy

While recognizing the importance of other factors, this paper focuses on the implications of the changing economy for science education. The approach used analyzes the views of educators, business leaders, economists, and governmental agencies that relate educational reform to America's need to compete effectively in a global economy, as well as our needs to be more productive, and to improve the quality of human capital. These economic changes are relevant to science education because the primary factors propelling them are advances in science and innovations in technology. In turn, these changes influence the ways that students are "civically prepared and economically empowered" (Carnegie Foundation for the Advancement of Teaching, 1988). The "knowledge, learning, information, and skilled intelligence (of the civically prepared and economically empowered person) are the raw materials of international commerce..." (The National Commission on Excellence in Education, 1983, p.71).

Our entrance into a knowledge-based global economy and the emergence of an information-based society have changed the culture of the workplace and the conceptions of the worker. This changing culture of work is viewed by many economists as resulting from a movement away from an industrialized society toward one dominated by services such as health, communications, transportation, finance, trade, education, real estate, and others (Guile & Quinn, 1988). In these areas, workers have more authority for decisionmaking, responsibility on the job, and responsibility for more diverse tasks.

Consequently, they must be flexible and capable of learning on the job because they are likely to work in teams rather than alone and generally must "work smarter" than in the past. These attributes have both social and technical implications. Levin (1988) finds that these trends extend to education as well. The teaching profession is newly characterized by initiative, cooperation, group work, peer training, evaluation, communication, reasoning, problem solving, decisionmaking, obtaining and using information, along with planning and learning skills. These new characteristics of teachers and teaching also have become goals of education.

As this example illustrates, business and education have a common interest in the implications of social and cultural changes for the welfare of the individual and the nation. However, neither group has discussed explicitly what these changes mean for the reform of science education. Business and education exist as separate cultures. This gap must be bridged if young people are to have opportunities for job security and a full life. Curricular reform which results from harmonizing science education with the economy will require collaborative efforts between business and the schools. The issue is not job training and vocational education, which are means. Instead, the issue is the goal: the education required in order to be able to live a full and productive life in a society where knowledge and change are the realities.

However, the schools' role in preparing youth for the workplace is subject to debate. For instance, potential philosophical and practical conflicts exist between the requirements of the economy and the purposes of schooling, between the requirements of the collective and the needs of the individual. Some scholars – Fritz Machlup, for instance – contend that no conflict exists between the economic requirements of the nation and the purposes of education. In the early 1960s, Machlup (1962), an economist, demonstrated the part that knowledge plays in determining an individual's worth as well as its significance in establishing the nation's gross national product. He pointed out that the amount of knowledge a person has really depends upon possessing skills for acquiring, coding, processing, and finally making use of information. His conclusions have particular implications in an economy where job opportunities are increasing for the "knowledgeable" and decreasing for the "nonknowledgeable." Machlup concluded that, under these conditions, no conflict exists between the economic requirements of the nation and the legitimate purposes of education.

Others, like T. W. Schultz, a Nobel Laureate in economics, concur. Schultz (1981) found that the acquisition of useful knowledge is a major factor in improving the quality of a population. Moreover, Boulding and Senesh (1983) argue that the betterment of the human condition is dependent upon the "optimal utilization of knowledge." Advances in scientific knowledge enhance not only the quality of both physical and human capital but are also the most powerful engine of production for the population as well as for the individual. In turn, the quality of production is dependent upon investments in education. The importance of science to the common good and to the improvement of the individual is underscored by these conclusions. The challenge for science education is to unlock knowledge in the sciences so that it becomes useful in individual affairs and serves the common good.

The Issue: Schooling and Work

While economic theorists demonstrate the importance of knowledge to productivity and to the betterment of the human condition, critics of education claim that the schools are not imparting the knowledge needed to perform adequately in the workplace. Much of the debate on the reform of science education has centered on the relationship of schooling to work. The President's Science Advisory Committee's Panel on Youth (1974) noted with concern that changes in social structures and occupational requirements have increased the number of jobs that require more education. However, the trend throughout this century has been to divorce work from schooling and the situation now appears to be "spinning out of control." The panel's perspective is that schooling should equip students "with cognitive and noncognitive skills relevant to their occupational futures, with knowledge of some portion of civilization's cultural heritage, and the disposition for acquiring such skills and knowledge" (p. 25). In addition, the panel agrees that business needs to create environments that complement the work of schools by providing opportunities for young people to experience the responsibilities of a productive worker in the "real world."

As economists Carnoy and Levin (1985) indicate, "Although the relationship between schooling and work is not direct, schooling is nonetheless defined by the workplace. The demand for better schooling is influenced by the knowledge requirements associated with different jobs and careers." The educational system is thus perceived as less and less successful in preparing the young for today's changing workplace.

Responses to the Issue of Schooling and Work

The preparation of youth for the changing workplace is a recurrent theme in the national reports that have deliberated on the problems of the changing economy and on the resulting imperatives for educational reform. Statements and conclusions from these reports have implications for school science. Excerpts from 26 reports appear in the Appendix. About half of the reports contain analyses of issues and recommendations relating to education and the economy from the educational community's perspective. The other reports contain views on the issues from the perspective of the private sector. Entries in the Appendix present that which, if anything, each report says about (a) conditions necessitating educational changes, (b) educational objectives, (c) curricular changes required, and (d) the coordination needed in the educational system to effect change.

Most significant is the consensus in the reports, which have been produced by different communities. Each of the reports, whether it originates from the business or the educational community, conveys a sense of urgency for the reform of school science. Each report expresses the belief that, because of changes in the culture and the economy augmented by advances in science and technology, the traditional science curricula are out of step with the world in which high school graduates will live, work, and assume the responsibilities expected of members of a modern society. Consequently, each report recognizes that the reconceptualization of science education is a matter of particular national concern.

Consensus also exists among the reports' goals for science education. The goals are to develop the abilities to (a) solve problems encountered in the workplace and in the conduct of personal life and civic responsibility, (b) communicate effectively, (c) work in groups, and (d) learn how to learn.

It is noteworthy that many of the issues raised in these reports have a long history. Some of the issues have been mentioned already—the appropriate and practical role of schools in promoting social changes and curing social ills, and maintaining the appropriate balance of the contributions of education to the good of the individual and the good of society. Another issue is illustrated in the writings of John Dewey (1940) who, a half-century ago, spoke of the need for schools to "develop an industrial intelligence" that would help people not only to solve problems at work but also to understand the scientific principles behind the transformation of work. Although the realities of life and work are different today, the challenge of educational reform remains the same: the development of science curricula that are in harmony with the complexities of living, learning, and working in modern America.

Reaching agreement on the essential components of scientific literacy necessary for understanding and working in modern America requires the cooperative efforts of those concerned with and informed about science education. This is a challenging task because so many different, isolated cultures are represented. Education and business, for instance, exist as separate cultures that typically have not interacted on the subject of science education.

In summary, the national reports on the state of education and the economy find existing science curricula socially, culturally, and cognitively outdated, thus placing "the nation at risk." With recent shifts in the economy, the workplace, social affairs, the ethos of science and technology, demographics, and the content of and instruction in school science today have become anachronisms. The central themes of the reform movement are to enable young people to comprehend the present and to plan the future. Thus, the focus of curricular reform must begin with the needs of the individual in a rapidly changing society. In this context, the slogan "scientific and technological literacy" implies a science education for understanding and working in the world in which we live.

The Next Step

As it stands now, modifications to the science curricula have been marginal. One senses a resistance to change among teachers because "authentic" guidance is lacking. Mandates and regulations by state and local political authorities have, for the most part, been counterproductive because they attempted to optimize traditional curricula. In their survey of business and school alliances, Atkin and Atkin (1989) found that "only infrequently are they designed either to add to the school's established curriculum or to offer a worthy departure from it" (p. 87). No longer can the nation afford to put another generation of young people at risk and to jeopardize the country's economic future by ignoring basic issues and merely tinkering with the structure of schooling. This is not the way to find a viable concept of scientific and technological literacy.

The time has come to bring together the people who are most informed about the natural, cognitive, and social sciences and about technology, education, and business to refine the reports that have been prepared already on aspects of the reform of science education. The recommendations in the national reports provide an agenda for the development of a detailed conceptual framework for modernizing the teaching of the sciences in our nation's schools. Each group represented will have the preliminary task of getting acquainted with the culture of the other groups. All of the groups will need to use a language that allows for the discussion of common educational concerns in an atmosphere of understanding and good will.

At present, there is a plethora of ambiguous goals and policy statements and of similarly vague descriptions of the qualities desired of workers in the new work force. Moreover, the rhetoric on the importance of educational reform is extensive, but systematic analysis of the issues is minimal. There is, however, as the Appendix demonstrates, some consensus about the directions that science education should take. Yet a common vision and a logically derived conceptual framework for new science curricula are lacking. Implementation of the framework will be another endeavor that will demand collaborative action and continuous support.

References

Atkin, J. M., & Atkin, A. (1989). *Improving science education through local alliances.* New York: Carnegie Corporation of New York. (Draft, mimeo). (p. 87).

Boulding, K. E., & Senesh, L. (Eds.). (1983). *The optimum utilization of knowledge: Making knowledge serve human betterment.* Boulder, CO: Westview Press.

Carnegie Foundation for the Advancement of Teaching. (1988). *An imperiled generation: Saving urban schools.* Princeton, NJ: Author.

Carnoy, M., & Levin, H. M. (1985). *Schooling and work in the democratic state.* Stanford, CA: Stanford University Press.

Dewey, J. (1940). *Education today.* New York: Putnam.

Guile, B. R., & Quinn, J. B. (Eds.). (1988). *Technology in services: Policies for growth, trade, and employment.* Washington, DC: National Academy Press.

Levin, H. M. (1988). *Economic trends shaping the future of teacher education.* Stanford, CA: Center for Education Research at Stanford (mimeo).

Machlup, F. (1962). *The production and distribution of knowledge in the United States.* Princeton, NJ: Princeton University Press.

National Commission on Excellence in Education. (1983). *A nation at risk: The imperative for educational reform.* Washington, DC: U.S. Government Printing Office.

Oxford Analytica. (1986). *America in perspective: Major trends in the United States through the 1990s.* Boston: Houghton Mifflin.

President's Science Advisory Committee, Panel on Youth. (1974). *Youth: Transition to adulthood.* Chicago: University of Chicago Press.

Schultz, T. W. (1981). *Investing in people: The economics of population quality.* Berkeley, CA: University of California Press.

Appendix

Excerpts from 26 Reports on the State of Education and the Economy

Views from the Educational Community

National Science Board Commission on Precollege Education in Mathematics, Science, and Technology

Educating Americans for the 21st Century: A Report to the American People and the National Science Board
(1983)

Conditions necessitating educational changes

The rapidly changing nature of the world has made the conventional teaching of science, mathematics, and technology inadequate for dealing with the problems of the future and for building the "basic capital of tomorrow's society."

Educational objectives

Subject matter should:

- Be applicable to the solution of practical problems
- Reflect the changing job markets
- Aid in dealing with scientific and social issues.

Students should be able to:

- Meet their civic responsibilities
- Cope with health problems
- Adapt to an increasingly technological world.

Students must learn to:

- Learn on their own
- Be willing to learn on their own
- Communicate effectively
- Reason logically
- Think critically, creatively, and innovatively.

Students must understand:

- How new knowledge is acquired
- How short- and long-term risks influence problemsolving and decisionmaking.

National Commission on Excellence in Education

A Nation at Risk: The Imperative for Educational Reform
(1983)

Conditions necessitating educational changes

We live in a new era in which "knowledge, learning, information, and skilled intelligence are the raw materials of commerce" (p. 7).

To maintain ever-renewable human resources will require a knowledge base and skills for lifelong learning.

Educational objectives

The knowledge base and skills for lifelong learning require that students be introduced to:

- The concepts, laws, and processes of the physical and biological sciences
- The methods of scientific inquiry and reasoning
- The application of scientific knowledge to everyday life
- The social and environmental implications of scientific and technological developments.

The Carnegie Forum on Education and Economy

A Nation Prepared: Teachers for the 21st Century
(1986)

Conditions necessitating educational changes

Structural changes have occurred in our economy, based on developments in science and technology, that make new demands upon individuals at every educational level and on those entering the work force.

Educational objectives

Transformation of the traditional school science curricula is a necessary step toward improvement of the educational system.

Curricular changes required

Qualitative and quantitative changes are needed to keep the teaching of science in line with social, economic, and individual affairs.

National Science Foundation

Human Talent for Competitiveness
(1987)

Conditions necessitating educational changes

A close relationship exists between the quality of life and new knowledge. The knowledge most prized in our society arises from advances in science and technology. This knowledge drives our economy.

Educational objectives

The science and engineering backgrounds of those entering the work force must be improved and increased.

National Governors' Association

Making America Work: Productive People, Productive Policies
(1987)

Conditions necessitating educational changes

Jobs, growth, and competitiveness are essential to "enhance our competitive position as a nation."

Educational objectives

- Workers must be prepared to anticipate change and to be more flexible in responding to it.
- The education and training of the work force must be lifelong endeavors.
- Students interested in science, mathematics, and technology must be provided with opportunities that encourage these interests.
- A greater understanding of international affairs, the development of a work ethic, and computer literacy must be imparted.

Robert B. Reich

Education and the Next Economy
(1988)

Conditions necessitating educational changes

In light of global economic changes, education for the future must be different from that of the present, especially if the nation is to become more competitive.

Educational objectives

The work force for a global economy must be:

- Innovative and able to transform ideas into incrementally better goods and services
- Able to make judgments and evaluations
- Adjustable and adaptable

- Able to generate and use information.

Levels of "numeracy," literacy, responsibility, and learning must be increased.

Students must possess good communication skills and the ability to collaborate with others.

Curricular changes required

Learning must be more interdisciplinary, flexible, and have more connections to the "real world" of the learner and society.

Coordination needed in the educational system

Styles of teaching modeled on the memorization of facts reinforced by drills, practice, and tests are viewed as more suitable for robots than for people.

Task Force on Women, Minorities, and the Handicapped in Science and Technology

Changing America: The New Face of Science and Engineering (1988)

Conditions necessitating educational changes

Cooperative efforts between schools, business, and colleges are needed in order to develop of a "fully competitive work force."

Educational objectives

- A knowledge of science and engineering is essential for all students if they are to reach their fullest potential as workers in contemporary America.
- Systematic changes are required in science and mathematics curricula, especially for grades K-8.

Curricular changes required

The new curriculum should emphasize:

- The practical application of technical theories
- That an understanding of science and technology is crucial to the workplace and the economy
- The influences of a knowledge of science for the conduct of personal affairs.

Office of Technology Assessment

Technology and the American Economic Transition: Choices for the Future (1988)

Conditions necessitating educational changes

Few jobs will remain unaffected by "new technologies, rapid increases in foreign trade, and the tastes and values of a new generation of Americans [which] are likely to reshape virtually every product, every service, and every job in the United States" (p. 3).

Educational objectives

Education must undergo changes in order to realize more productive learning and learners capable of greater self- direction. Students must acquire:

- The ability to adapt quickly
- Skills for coping with ambiguity and with too much or too little information
- Basic literacy in mathematics, science, and information technology
- The ability to work cooperatively in groups.

Curricular changes required

The gaps between formal and abstract learning and their practical or real world applications must be bridged.

William T. Grant Foundation Commission on Work, Family, and Citizenship

The Forgotten Half: Pathways to Success for America's Youth and Young Families (1988)

Conditions necessitating educational changes

Effective learning is closely related to successful performance in the workplace and, therefore, to the nation's productivity.

Educational objectives

The educational program should emphasize:

- Lifelong learning
- Cooperative learning strategies

- Instructional processes that require critical thinking and problemsolving.

Curricular changes required

Achieving a more productive education will require

- That students have a mixture of abstract and experiential learning
- The combination of conceptual study with concrete applications.

Coordination needed in the educational system

- Knowledge should be integrated into action.
- Learning should be coordinated with community-based activities.

American Association for the Advancement of Science

Science for All Americans—A Project 2061 Report on Literacy Goals in Science, Mathematics, and Technology
(1989)

Conditions necessitating educational changes

Significant discoveries and insights in the sciences, along with innovations in technology and the impact both have on the lives of people, occur at a pace that suggests we ought to rethink the rationale for the teaching of science three or four times per century.

Educational objectives

Science and technology must be made more intellectually vigorous.

Curricular changes required

- The volume of subject matter typical of traditional courses needs to be reduced selectively.
- Science should be taught in a societal context, with attention to practical applications.
- The connections among the sciences, mathematics, history, literature, and other school subjects should be emphasized.
- An understanding of the meaning of scientific inquiry should be fostered.

Criteria for the selection of subject matter include:

- Recognition of economic factors and their meaning for work and productivity
- Recognition of the significance of science for everyday living and for responsible social participation
- Attributes that will contribute to a satisfactory quality of life.

National Center on Education and the Economy

To Secure Our Future: The Federal Role in Education
(1989)

Conditions necessitating educational changes

The competitive strength of the nation arises from an opportunity for all children to develop to their fullest capacities. We need to restructure elementary and secondary education "to secure our future."

Educational objectives

The major goals of schooling are to develop:

- Higher-order thinking skills
- The ability to use knowledge to solve complex problems.

Curricular changes required

Efforts must be made to stop compounding learning failures.

Coordination needed in the educational system

- The integration of social services programs with the work of schools is encouraged.
- Business has a responsibility to share in educational reform by establishing communication networks between schools and business for the development of programs focused on the transition from school to work.

Education Commission of the States, Task Force on Education for Economic Growth

Action for Excellence: A Comprehensive Plan to Improve Our Nation's Schools (1983)

Conditions necessitating educational changes

The development of a highly competitive global economy, stimulated by advances in science and technology, forces a reconsideration of our nation's educational goals.

Educational objectives

Well-rounded students must know how to:

- Learn
- Apply knowledge
- Use modern informational technologies
- Improve reading, writing, and mathematical skills
- Recognize valid concepts
- Formulate problems, find ways of solving them, and defend the rationality of conclusions
- Reason inductively and deductively
- Use information presented in various forms such as spoken, written, tabular, or graphic
- Adapt to change
- Relate knowledge and skills to employability and promotability.

Coordination needed in the educational system

Cooperation among teachers, business and labor leaders, and educators is essential. All levels of schooling must be involved, from kindergarten through college.

The Private Sector – Business and Industry

Susan W. Sherman
Education for Tomorrow's Jobs
(1983)

Conditions necessitating educational changes

Economic competitiveness among nations and the development of a world economy have changed the pattern of employment opportunities for today's youth. We need "a new perspective to the relationships among vocational education, economic development, and the private sector" (p. v).

Educational objectives

All education should be viewed as having a vocational component in that the skills most essential to work are the abilities to read, write, speak, reason, and compute.

Coordination needed in the educational system

Collaborative efforts between schools and business are seen as indispensable for the improvement of schooling.

James B. Hamilton
The Public Reacts to "Education for Tomorrow's Jobs"
(1984)

Educational objectives

Students need to receive training in:

- how to learn
- how to analyze situations
- how to solve problems.

Required curricular changes

- General courses should be integrated with real-life activities.
- General and vocational education should center around a broad-based core curriculum studied by all students.

Center for Public Resources

Basic Skills in the U.S. Work Force: The Contrasting Perceptions of Business, Labor, and Public Education
(1983)

Conditions necessitating educational changes

Business leaders regard precollege education as an investment in human capital to ensure work opportunities for the individual and the economic growth of communities and the nation.

Traditionally, schools have educated students for their first job only, rather than for a 50-year investment in progressive job advancements.

Educational objectives

School and business ventures need to pay more attention to the development of "basic skills."

Business seeks a range of work competencies:

- Writing
- Reading
- Speaking
- Listening
- Mathematical, scientific, and technological reasoning abilities.

National Academy of Sciences, National Academy of Engineering, Institute of Medicine

High Schools and the Changing Workplace: The Employers' View
(1984)

Conditions necessitating educational changes

The U.S. economy depends on high school graduates who have not attended college but who make up the major portion of the work force.

Educational objectives

"A person who knows how to learn is one well grounded in fundamental knowledge and who has mastered concepts and skills that create an intellectual framework into which new knowledge can be added" (p. 17).

The changing nature of the workplace requires that students have the ability to:

- Learn on their own

- Acquire new knowledge continually
- Adapt to change
- Offer constructive dissent without hindering team work.

Elements of this "intellectual framework" include:

- Numerical and reading skills
- The ability to gather information and organize it in a logical and coherent manner
- The ability to interpret, draw inferences, and communicate information effectively
- Capacities for reasoning, problemsolving, and decisionmaking
- An understanding of specific concepts in the biological and physical sciences as well as in technology
- The ability to work cooperatively with others.

Business Advisory Commission of the Education Commission of the States

Reconnecting Youth – The Next Stage of Reform
(1985)

Conditions necessitating educational changes

There is a need for business and education to collaborate in establishing policies that can improve the transition of young people to become productive members of the work force.

This is a critical problem for at-risk youth and entails specific actions to connect them with society and with work.

Curricular changes required

Schools are encouraged to:

- Abolish tracking
- Mix vocational and general education students
- Foster peer teaching and collaborative learning projects
- Consider magnet schools and classes.

Coordination needed in the educational system

More cooperative efforts are necessary between schools and business.

Chamber of Commerce of the United States

Business and Education: Partners for the Future
(1985)

Conditions necessitating educational changes

Changing job patterns in the United States call for new and higher levels of skills, literacy, and training for those entering the work force.

Educational objectives

All students should be enrolled in a "basic core curriculum" that is tailored to:

- An information society
- The stresses of lifelong learning
- A life of continuous job updating.

Curricular changes required

The school curriculum should:

- Be more responsive to the job market, especially in the sciences and mathematics
- Recognize the impact of rapidly paced high technology and a competitive economy.

Coordination needed in the educational system

A "school-to-work" program is advocated but not defined.

Committee for Economic Development

Investing in Our Children: Business and the Public Schools
(1985)

Conditions necessitating educational changes

We need to redefine the intellectual and behavioral traits that are essential for employability and success in the work force.

Educational objectives

The goals should be to:

- Provide greater opportunities and a more productive life for all children

- Assure economic growth for the nation
- Improve social well-being.

Curricular changes required

Achievement of these ends requires an education that stresses:

- Literacy
- Mathematics
- Problemsolving
- Learning skills
- The ability to apply knowledge
- The capacity to adapt to change
- Team work
- The capacity to assume responsibility
- Self discipline
- Respect for the rights of others.

National Commission on Secondary Vocational Education in the High School

The Unfinished Agenda: The Role of Vocational Education in the High School (1985)

Educational objectives

A more balanced educational program is needed in secondary schools, one that recognizes students' diversities in ability and interest.

Curricular changes required

Vocational education programs should stress:

- Problemsolving, analytical thinking, scientific inquiry, and reasoning
- Basic communication and interpersonal skills and promote their transferability to various settings
- The ability to gather and analyze information and to recognize the interrelation of ideas
- An appreciation of the implications of technological development

- An understanding of the fundamentals of how our economic system works.

Richard M. Cyert and D. C. Mowery

Technology and Employment: Innovation and Growth in the U.S. Economy
(1987)

Conditions necessitating educational changes

Technological change is an essential component of a dynamic, expanding economy. One result of this change is to cause individuals to make "painful and costly adjustments" as new technologies are adopted, or face unemployment. Whether a worker becomes unemployed by technological changes depends in part on "the amount and quality of basic skills preparation provided to labor force entrants by U.S. public schools" (p. 11).

Educational objectives

The essential skills needed for employment are:

- Problemsolving
- Numerical reasoning
- Written communication
- Basic literacy.

W. B. Johnston and A. H. Packer

Workforce 2000: Work and Workers for the 21st Century
(1987)

Conditions necessitating educational changes

The economic assets of a nation today are represented in the knowledge, skills, and motivations of people – no longer in the natural resources or industrial plants.

Building human capital is dependent upon appropriate education and training and the trend is for ever-increasing amounts if the economy is to grow.

Educational objectives

By the year 2000, the secondary school graduate should be proficient in:

- Reading sophisticated materials
- Writing clearly
- Speaking articulately

- Solving complex numerical problems.

Coordination needed in the educational system

Studies should be undertaken to analyze the interconnections among employment, education, literacy, cultural values, income, and living environments.

National Governors' Association and the Conference Board

The Role of Science and Technology in Economic Competitiveness
(1987)

Conditions necessitating educational changes

The most demanding challenge facing America's leadership today is to restore the country to its competitive position in the global marketplace.

Education is the key to building the human resources essential to the restoration of the nation's competitive strength.

Educational objectives

- Education needs new goals which stress lifelong learning.
- Students should no longer expect that work skills learned in school will last a lifetime.
- The work force for a changing economy must be flexible.

Coordination needed in the educational system

Meeting the educational requirements of the new work force depends in part on the identification by business of specifications and long-term perspectives for the development of human resources.

Office of Technology Assessment

Educating Scientists and Engineers: Grade School to Grad School
(1988)

Conditions necessitating educational changes

There is a continuous and dynamic movement in the sciences and in technology which "makes some disciplines obsolete while creating new ones."

Advances in communication and information technologies are particularly significant factors.

Educational objectives

Elementary and secondary school education needs to be restructured as a whole.

Coordination needed in the educational system

In suggesting that schools should "do better," the curriculum and teacher education were targeted as critical areas. However, there were no specific suggestions of what the nature of these improvements should be.

Office of Public Affairs, Employment and Training Administration, U.S. Department of Labor

Building a Quality Workforce: A Joint Initiative of the U.S. Departments of Labor, Education, and Commerce
(1988)

Conditions necessitating educational changes

The economy and the workplace are changing, creating a growing gap between schooling and job requirements in the work force. This situation threatens "our nation's economic strength and vitality, and our international competitiveness depend on our capacity to build and maintain a quality work force" (p. 1).

Educational objectives

The trend toward more flexible modes of working and the decentralization of authority demand workers with abilities in:

- Problemsolving
- Team work
- Initiative
- Self direction
- Adaptability to change
- Lifelong learning skills and processes.

Critical skills include:

- Reading
- Writing
- Mathematics
- Communication.

American Society for Training and Development
Workplace Basics: The Skills Employers Want
(1988)

Conditions necessitating educational changes

"The workplace is changing and so are the skills that employees must have in order to change with it" (p. ii).

Educational objectives

The nature of economic changes and heightened competitiveness both require an "upskilling" of workers.

Workers must:

- Be more adaptive
- Be innovative
- Possess strong interpersonal skills
- Be able to work as team members.

Workers must have the ability to:

- Set goals
- Solve problems
- Make decisions
- Be good listeners
- Take responsibility
- Recognize when, where, and how they should assume a leadership role.

To adapt quickly as the workplace changes, the worker must:

- Know how to learn
- Think creatively
- Possess essential mathematics and reading skills.

Coordination needed in the educational system

The knowledge and learning activities presented in schools must be integrated with those required in business.

3

Scientific Illiteracy: Causes, Costs, and Cures

John Bishop

A large gap exists between the science and mathematics achievements of American high school students and those of their counterparts overseas. Few American students are exposed to rigorous science courses in high school and the achievement levels of the students in these courses rank below those of students in most other industrialized nations (see Appendix A). In physics, for example, the 11% of Australian and 24% of Norwegian 18-year-olds studying the subject in the final year of secondary school did better on the international physics test than the very select 1% of American seniors taking their second year of physics (most of whom were in Advanced Placement courses).

This pattern of poor achievement in science subjects by American high school students provides vivid evidence of the scientific and mathematical deficiency of the population entering this country's work force directly from secondary school. Leaders from every sector of society are calling for improvements in the teaching of science, mathematics, and technology as means of improving the quality of the labor force and the quantity of the nation's economic productivity. While it is true that major conceptual changes are needed in the ways that science, mathematics, and technology are taught, these changes will have only a slight impact unless the incentives and rewards for studying science, mathematics, and technology are altered radically.

This chapter examines the causes of the well-documented learning deficits of American youth in science, mathematics, and technology, evaluates their social costs, and recommends policy measures for remedying the problems identified. Following the lead of the American Association for the Advancement of Science's *Science for All Americans* (1989) report, I define the domain of "science" very broadly to include mathematics and technology along with the natural sciences. To avoid confusing readers accustomed to the narrower definition of science, the phrase "science" is used herein when referring to science as defined in the previous sentence.

The paper is organized into four sections. Section 1 presents evidence showing that American students devote considerably less time and energy to studying science, mathematics, and technology in high school than do their counterparts abroad.

Section 2 traces the apathy of students and their parents toward science, mathematics, and technology education to the failure of our society to recognize and reward those students who are willing to commit the time and energy necessary to master these subjects. Competence in science has *no* impact on the wage rates or earnings of people under the age of 30 and very little impact on their chances of admission to the most prestigious colleges. Competence in mathematical reasoning has *no* effect on the success in the labor market of young men and *only very limited* effect on success of young women. Proficiency in technology, by contrast, has *very substantial positive* results on the wage rates and earnings of young males.

Section 3 examines the social costs of the learning deficits in science, mathematics, and technology. Mathematical and technological skills are found to have substantial effects on the hands-on measures of job performance in military jobs that are similar to the blue-collar and technical jobs occupied by the majority of male high school graduates. Proficiency in mathematical reasoning has substantial effects on job performance in clerical jobs. Scientific knowledge has modest effects on job performance in both types of work. This implies that, even though the labor market fails to reward most young people who have studied and learned science and mathematics, the nation's future productivity will be increased if high school students receive a stronger grounding in mathematics and science, although the results of scientific knowledge are not as large as one might hope. For young men and women hoping to obtain blue-collar and technical jobs, an empirical analysis suggests that technological competence is a powerful contributor to both productivity and success in the labor market. The *Nation at Risk* (1983) report's recommendation that every student take a course in computers gave some recognition to the need for technology education, but computers are only one of the technologies with which we interact on a daily basis. This is an area of study that needs much more attention than it has been receiving.

Section 4 sets forth a series of policy recommendations designed to improve incentives for students to devote more time and energy to certain types of learning and to strengthen parental incentives to demand that the performance of local schools be upgraded.

Apathy: The Proximate Cause of the Science Learning Deficit

The primary reason American high school students do poorly in international comparisons is that they devote a lot less time and energy to learning than do their counterparts in other countries. American students average nearly 20 absences a year; Japanese students only 3 a year (Berlin & Sum, 1988). School years are longer in Europe and Japan. Thomas Rohlen has estimated that Japanese high school graduates spend on average the equivalent of three more years in school and in studying than do American high school graduates (Quinlan & Merrow, 1989). Studies of time use and time on task show that American students are engaged attentively in learning activities for only about half of the time they are in school. A study of schools in Chicago found that, in public schools with high-achieving students, an average of about 75% of the time was spent on actual instruction; in schools with low-achieving students, the average was 51% (Frederick, 1977). Overall, Frederick, Walberg, and Rasher (1979) estimated that 46.5% of the potential learning time was lost due to absence, tardiness, and inattention.

In the *High School and Beyond Survey,* students reported spending an average of 3.5 hours per week on homework (National Opinion Research Corporation, 1982, Q. 15). When homework is added to engaged time at school, the total time devoted to study, instruction, and practice is only 18 to 22 hours per week – between 15 and 20% of the student's waking hours during the school year. By way of comparison, the typical senior spent 10 hours per week in a part-time job and about 24 hours per week watching television (A. C. Nielsen, 1987). Thus, watching television occupies as much time as formal learning. Students in other nations spend much less time watching television: 55% less in Finland, 70% less in Norway, and 44% less in Canada (Organization for Economic Cooperation and Development, Table 18.1, 1986). Science and mathematics' deficits are particularly severe because most American students do not take demanding college preparatory courses in these subjects. The high school graduating class of 1982 took an average of only 0.43 credits of algebra II, 0.31 credits of more advanced mathematics courses, 0.40 credits of chemistry and 0.19 credits of physics (Meyer, 1988, Table A.2).

Even more important than the time devoted to learning is the intensity of the student's involvement in the process. At the completion of his study of American high schools, Theodore Sizer (1984) characterized students as *"All too often docile, compliant, and without initiative"* (p. 54). John Goodlad (1983) described *"a general picture of considerable passivity among students..."* (p. 113). Sixty-two percent of 10th graders agree with the statement: "I don't like to do any more school work than I have to" (Public Opinion Laboratory, 1988, Q. AA37N). The high school teachers surveyed by Goodlad ranked "lack of student interest" and "lack of parental interest" as the two most important problems in education.

The students' lack of interest makes it difficult for teachers to be demanding. Sizer's description of Ms. Shiffe's class illustrates what happens sometimes:

> Even while the names of living things poured out of Shiffe's lecture, no one was taking notes. She wanted the students to know these names. They did not want to know them and were not going to learn them. Apparently no outside threat – flunking, for example – affected the students. Shiffe did her thing, the students chattered on, even in the presence of a visitor....Their common front of uninterest probably made examinations moot. Shiffe could not flunk them all, and, if their performance was uniformly shoddy, she would have to pass them all. Her desperation was as obvious as the students' cruelty toward her (Sizer, 1984, pp. 157–158).

Some teachers are able to overcome the obstacles and persuade their students to undertake difficult learning tasks. But, for most, the students' lassitude is demoralizing. Teachers are assigned responsibility for setting high standards, but they are not given the tools that might be effective for inducing student observance of the academic goals of the classroom. Ultimately, teachers must rely on the force of their own personalities. All too often, teachers compromise academic demands because most of their students see no need to accept them as reasonable and legitimate.

The Apathy of Parents and School Boards

The second major reason for the low levels of science and mathematics achievement is apathy on the part of parents and school boards. A National Science Foundation-funded survey of 2,222 parents of 10th graders found that 25% thought their child should take only one or two science classes in high school (Public Opinion Laboratory, 1988, Q. BH165). When 2,829 high school sophomores were asked whether "My parents...think that math [*sic*] (science) is a very important subject," 40% said "no" with respect to mathematics and 57% said "no" for science (Public Opinion Laboratory, 1988, Q. AA19Q–AA19R). Only 30% of the 10th graders reported that their parents "want me to learn about computers" (Public Opinion Laboratory, 1988, Q. AA19D).

Japanese families allocate 10% of the family's income to educational expenses, American families, only 2%. If American parents were truly dissatisfied with the performance of their local public schools, they could send their children to tuition-financed schools offering enriched and rigorous curricula (as many Australian parents do) or arrange for their children to be tutored (as half of the Japanese parents do). Private investment in secondary education may be relatively low in the United States because parents are satisfied with the education their children are getting in the public schools.

A comparative study of primary education in Taiwan, Japan, and the United States found that *even though American children are far behind Taiwanese and Japanese children in mathematics capability, American mothers are much more pleased with the performance of their local schools than Taiwanese and Japanese mothers are.* When asked "How good a job would you say ___'s school is doing this year educating ___," 91% of the American mothers responded "excellent" or "good" while only 42% of the Taiwanese and 39% of the Japanese parents were this positive (Stevenson, 1983). Clearly, American parents hold their children and their schools to lower academic standards than do Japanese, Taiwanese, and European parents. Why is this the case?

Incentives: The Real Cause of the Science Learning Deficit

Incentives for Learning Science and Mathematics in High School

The fundamental causes of student and parental apathy toward science, mathematics, and technology education are the absence of reliable indicators of science and mathematics learning in high school and the consequent lack of rewards for learning science and mathematics. The indicators of an individual's learning generated by our educational system, such as years of schooling and scores on the Scholastic Aptitude Test, generate handsome rewards —better-paying jobs and admission to prestigious colleges. By contrast, the indicators of science and technology learning accomplishments in high school that are communicated to colleges and employers are weak. Learning accomplishments for which good learning indicators do not exist are not rewarded.

The lack of incentives for learning science, mathematics, and technology is a consequence of three phenomena:

- The peer group actively discourages academic effort.
- Admission to the most desirable colleges is not influenced significantly by achievements in science and technology. Instead, it is based on aptitude tests, which do not assess competence in science, and on class rank and grade point averages, which are defined relative to classmates' performances, not relative to an external standard.
- The labor market does not reward science and mathematics achievement in high school.

The Zero-Sum Nature of Academic Competition in High School

An important cause of high school students' poor motivation is peer pressure against studying hard. Students who enjoy science and work hard in science courses are considered to be "nerds" by many of their classmates. A primary reason for peer pressure against studying hard is that pursuing academic success forces students into a zero-sum competition with their classmates. Their achievement is not measured against an objective, external standard. By contrast to scout merit badges, for example, where recognition is given for attaining a fixed standard of competence, the schools' measures of achievement assess performance relative to fellow students through grades and class rank. When students try hard to excel, they set themselves apart, cause rivalries, and may make things *worse* for their friends. *When we set up a zero-sum competition among close friends, we should not be surprised when they decide not to compete.* All work groups have ways of sanctioning "rate busters." High school students call them "brain geeks," "grade grubbers," and "brown nosers."

The second reason for peer norms against studying is that most students perceive the chance of receiving recognition for an academic achievement to be so improbable that they give up trying. At most high school awards ceremonies, academic recognition goes to only a few—those at the very top of the class. By ninth grade, most students are already so far behind the leaders that they know they have no chance of being successful academically. Often, their reaction is to dismiss the students who take learning seriously and to honor other forms of achievement—athletics, dating, holding their liquor, and being "cool"—which are more exciting and offer them better chances of acceptance and social success.

College Selection Criteria

In Canada, Australia, Japan, and Europe, educational systems administer achievement examinations in science, mathematics, and other subjects that are closely tied to the curriculum. With the exception of Japan, all of these examinations use an extended answer format. Performance on these examinations is the primary determinant of admission to a university and to a field of study and good grades on the most difficult examinations—physics, chemistry, and advanced mathematics—carry particular weight. In the United States, by contrast, the national tests that influence college admission decisions—the Scholastic Aptitude Test and the American College Testing Program—are multiple choice examinations that do not assess students' knowledge and understanding of science and technology.[1] The American examinations that are similar to those administered in Canada, Australia, and Europe—the Advanced Placement examinations in calculus, sciences, languages, and history—are taken by only 6.6% of high school seniors and have little impact on college admission decisions.

High school grade point averages and class rankings have substantial effects on who is admitted to the most prestigious colleges. Because most classes are graded on a curve, *taking more rigorous science and mathematics courses lowers students' grade point averages*.[2] Many college admission officers try to factor course difficulty into their evaluations, but most high school students still believe that A's in regular classes are better than B's in honors classes. The result is that many students avoid taking the more demanding courses such as chemistry, physics, and calculus.

The Absence of Major Economic Rewards for Effort in High School

Students who plan to look for a job immediately after leaving high school typically spend considerably less time studying science and mathematics than do those who plan to attend college. In large part, most of them see very little connection between how much they learn and how much they earn. When 10th graders are asked: "Which of the following mathematics [and science] courses will you need to qualify for your first choice of job," only 23% check geometry, 29% check algebra, 18% check trigonometry, and only 20 to 21% check biology, chemistry, and physics (Public Opinion Laboratory, 1988, Q. BA24B–BA25D). Statistical studies of the youth labor market confirm students' skepticism about the benefits of taking difficult mathematics and science courses and studying hard.

A study of 1972 high school graduates by Joseph Altonji (1988) for the National Center for Education and Employment found that, when family background and years of schooling are taken into account, the number of science courses taken in high school had no effect on wage rates in the first 14 years after graduation from high school. Science courses were associated with higher rates of wage growth, however, so there may be more substantial benefits later, when the individuals reach the age of 40. By contrast, mathematics courses and industrial, trade, and technical courses had significant positive results. The consequences of two additional, full-year mathematics courses on wage rates was between 0.88% and 3.4%, depending on specification. Two, full-year courses in an applied technology field raised wage rates by 2.5 to 2.8%.

The results of an analysis of the ability of scores of the Armed Services Vocational Aptitude Battery subtests (sample questions for each subtest are provided in Appendix B) to predict the labor market success of men and women in the youth cohort of the National Longitudinal Survey are summarized in Figures 1 and 2 (Bishop, 1988b). *It was found that, holding years of completed schooling and college attendance constant, young men received no rewards from the labor market for developing competencies in science, language arts, and mathematical reasoning (the ability to do arithmetic word problems and mathematical knowledge—knowledge of algebra and geometry, "high school" mathematics) during the first 8 years after leaving high school.* The only competencies that were rewarded were speed in doing simple computations (which calculators do more efficiently than people) and technical competence (knowledge of mechanical principles, electronics, automobiles, and shop tools). For the noncollege-bound female, there were both wage rate and earnings' benefits to learning advanced mathematics, but no benefits to developing competencies in science or technical areas. Competence in language arts did not

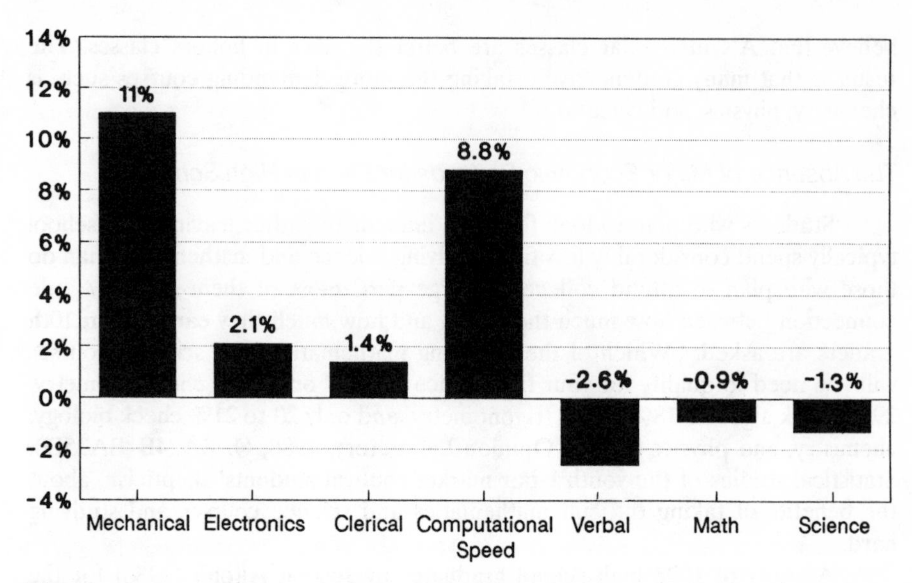

Figure 1a
Effect of Competencies on Earnings of Young Men 1984–1985

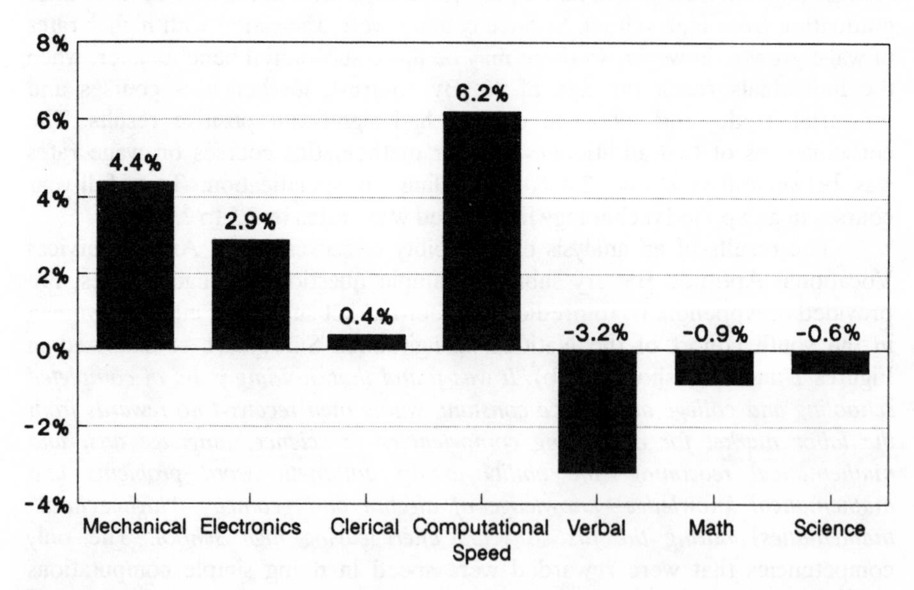

Figure 1b
Effect of Competencies on Wage Rates of Young Men 1983–1986

Source: Analysis of National Longitudinal Survey youth data. The figure reports the effect of a one-population, standard deviation increase in Armed Services Vocational Aptitude Battery subtest while controlling for schooling, school attendance, age, work experience, region, Standard Metropolitan Statistical Area residence, and ethnicity.

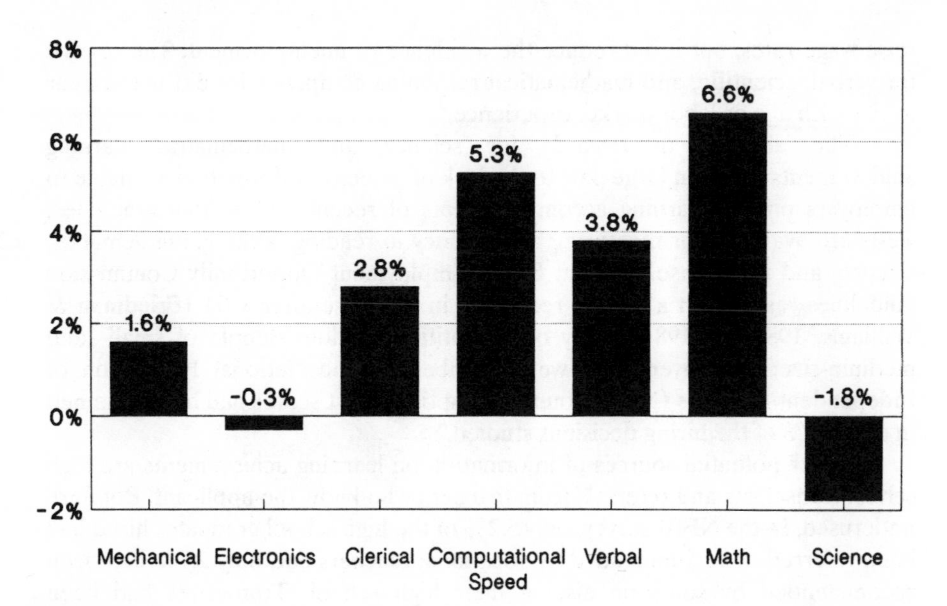

Figure 2a
Effect of Competencies on Earnings of Young Women 1984–1985

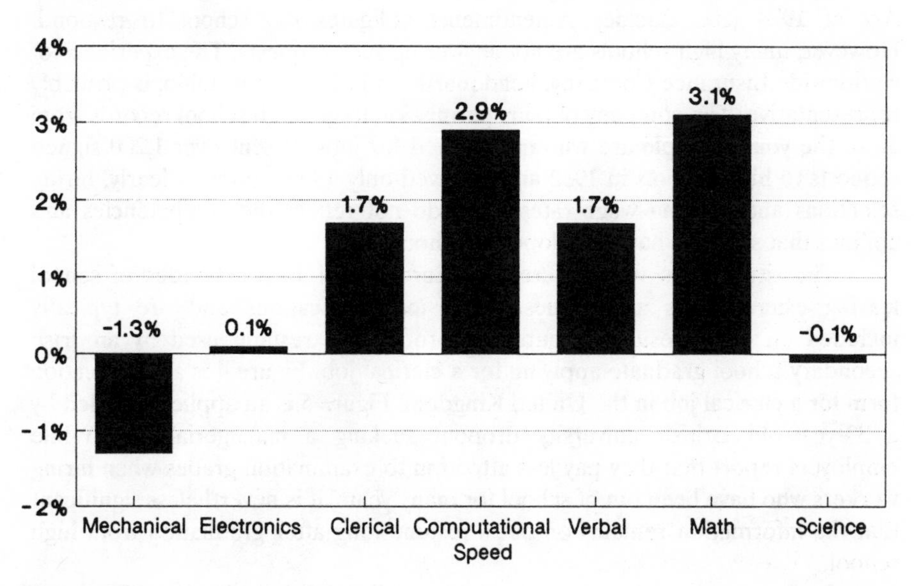

Figure 2b
Effect of Competencies on Wage Rates of Young Women 1983–1986

Source: Analysis of National Longitudinal Survey youth data. The figure reports the effect of a one-population, standard deviation increase in Armed Services Vocational Aptitude Battery subtest while controlling for schooling, school attendance, age, work experience, region, Standard Metropolitan Statistical Area residence, and ethnicity.

raise wage rates, but it did reduce the incidence of unemployment. The reward for verbal, scientific, and mathematical reasoning competencies did not appear to rise with age or labor market experience.[3]

The absence of rewards for science and mathematics learning achievements is due in large part to the lack of objective information available to employers on the learning accomplishments of recent high school graduates. Tests are available for measuring competency in reading, writing, mathematics, science, and problemsolving, but Equal Employment Opportunity Commission guidelines resulted in a drastic reduction in their use after 1971 (Friedman & Williams, 1982). A 1987 survey of a stratified random sample of small- and medium-sized employers who were members of the National Federation of Independent Business (NFIB) found that aptitude test scores had been obtained in only 2.9% of the hiring decisions studied.[4]

Other potential sources of information on learning achievements are high school transcripts and referrals from teachers who know the applicant. Both are underused. In the NFIB survey, only 5.2% of the high school graduates hired had been referred or recommended by vocational teachers and only 2.7% had been recommended by someone else in their high school. Transcripts had been obtained prior to the selection decision for only 14.2% of the high school graduates hired. If a student or graduate has given written permission for a transcript to be sent to an employer, the Family Educational Rights and Privacy Act of 1974 (the Buckley Amendment) obligates the school to respond. However, many high schools are not answering such requests. The experience of Nationwide Insurance Company, headquartered in Columbus, Ohio, is probably representative. The company obtains permission to get high school records from all of the young people are who interviewed for jobs. It sent over 1,200 signed requests to high schools in 1982 and received only 93 responses. Clearly, hiring selections and starting wage rates often do not reflect the competencies and abilities that students have developed in school.

The situation is very different in Europe and Japan. Grades of school leaving examinations are requested on job applications and are typically included on one's resume. Figure 3 reproduces a resume used by an Irish secondary school graduate applying for a clerical job. Figure 4 is an application form for a clerical job in the United Kingdom. Figure 5 is an application filed by a 33-year-old British university dropout seeking a managerial job. While employers report that they pay less attention to examination grades when hiring workers who have been out of school for many years, it is nevertheless significant that the information remains on one's resume long after graduation from high school.

In Japan, clerical, service, and blue-collar jobs at the best firms are available only to those recommended by their high school. The most prestigious firms have long-term arrangements with particular high schools to which they essentially delegate the responsibility of selecting new employee(s) for the firm.

CURRICULUM VITAE

Name ;
Address ;
Date of Birth ;
Place of Birth ;
Nationality ;
Marital Status ;
Occupation ;
Father's name ;
Occupation ;

EDUCATION :

August 1980 – June 1985.
(All five years were spent learning through Irish.)

U.C.D.

QUALIFICATIONS :

Subject:	Intermediate Cert. (June 1983)	Leaving Cert. (June 1985)
Irish	C (H)	C (H)
English	D (H)	C (H)
Mathematics	B (H)	A (P)
French	D	C (P)
German	D	–
Science	A	–
Chemistry	–	C (H)
Physics	–	C (H)
History	B	C (H)
Geography	B	–

Figure 3
Reproduction of an Irish Secondary School Graduate's Resume Submitted for a Clerical Position

APPLICATION FORM A-F CONFIDENTIAL	
POSITION APPLIED FOR	**RETURN TO**
Surname	Permanent Address
Forename(s)	
Date of Birth	
Place of Birth	
Nationality	Telephone No.
Marital Status	Next of Kin
No. of Children	Relationship
Sex and Year of Birth of Children (under 18 yrs)	Address
Do you own a car or motorcycle?	Do you hold a current driving licence?

MEDICAL HISTORY
Please state any serious illness, disability, allergy or operation..........
Are you registered as disabled?.......... If yes, state number..........

EDUCATION since the age of 11yrs

School	Dates From	To	Qualification	Subject and Grade

FURTHER EDUCATION –Give details of any qualifications gained since leaving school.

College/Evening Classes etc.	Dates From	To	Qualification	Subject and Grade

Institute(s) Membership & Grade

In what Language(s) other than English can you conduct business?

Figure 4
Application for a Clerical Job in the United Kingdom

APPLICATION FOR APPOINTMENT

Appointment applied for. Distribution Projects Manager (B+Q) Ref.No. ....

PERSONAL DETAILS: (block capitals)

Surname .... Title Mr. Forenames Mervyn John

Address ....

.... Postal Code .... Tel.No.Home .... Work ....

Marital Status M Children/Dependants (with ages) 1 x 4yrs., 1 x 1yr

Age 33 Date of Birth 5-8-55 Nationality British Place of Birth ....

State of health OK Height 6' Weight 13st 12lbs

Any disabilities/recurrent medical problems? None Regd.disabled —

Driving Licenses Car Car Owner ✓ Company Car —

Endorsements, convictions, accidents, etc. None

Leisure activities and offices held in clubs and societies Cycling/Walking

....

EDUCATION:

Secondary Education

From	To	School	Exams Taken (inc. grades)	Other achievements
1966	1972	Barnstaple Grammar	'O' Level: Eng. Lang. (2), Maths (3), French (3), Geog. (3), Statistics (3), Chemistry (3), Addl. Maths (6), Hist. (6), Physics (6) 'A' Level: Chem. (E), Physics (E), Maths (O)	Middle Sch. Games Captain

Further Education

From	To	College/Univ.	Course & results (inc.class/grades)	Other achievements
1972	1973	Univ. of Bradford	Applied Chemistry - left after 1 year - domestic reasons	

Other training and qualifications (inc. in-company and external courses, etc.)

From	To	Establishment	Training/Qualifications
1979		Farnley College, Leeds	Cert. of Prof. Competence (transport ops.)
1983	1984	Bradford College	Inst. of Industrial Management Cert.
1984	1989	In-Company	Numerous Management Classes

Membership of professional bodies

Date	Association/Institute	Grade of membership	Offices held

Figure 5

Application Form Completed by an Older University Dropout for a Managerial Position in the Private Sector in the United Kingdom

The criteria by which the high school makes its selection are: mutual agreement, grades, and examination results (Rosenbaum & Kariya, 1987).

Because information on school performance is available to employers, recent secondary school graduates in these countries are hired into primary labor market jobs that offer considerable training and access to career ladders at high-paying firms. In the United States, recent high school graduates are never even considered for these high quality jobs.

Incentives to Upgrade Local Schools

The lack of external standards for judging learning achievement in science, the resulting zero-sum nature of academic competition in school, and the lack of economic rewards for science learning also influence parents, school boards, and local school administrators. Parents can see that, on average, higher standards in science classes or hiring better science teachers will not, on average, improve their child's rank in class or grade point average. Generally, parents believe that improving the teaching of science at the local high school will have only minor effects on how their child does on the Scholastic Aptitude Test, so why should they worry about standards? Scores on Advanced Placement examinations and other achievement tests have little bearing on admission to better colleges.

The parents of children not planning to go to college have even less incentive to demand high standards at their high school. They believe that what counts in the labor market is having a high school diploma, not learning chemistry. They can see that learning more will be of only modest benefit to their child's future, and that higher standards might put at risk what is really important – the diploma.

In our system, locally elected or appointed school boards and the administrative and teaching staffs hired by these boards make the decisions – on teacher salaries, teacher hiring, the availability of Advanced Placement courses, grading standards, homework assignments – that determine the quality of education in local schools. If parents do not generate substantial grass-roots pressure for higher standards, state mandates designed to upgrade the quality of instruction will not have a lasting impact.

The Social Costs of the Learning Deficit

Will the learning deficit in science, mathematics, and technology have major consequences for the nation's standard of living? In the view of the National Commission on Excellence in Education:

> If only to keep and improve on the slim competitive edge we still retain in world markets, we must dedicate ourselves to the reform of our educational system.... Learning is the indispensable investment

required for success in the information age we are entering (1983, p. 7).

Behind the call for higher standards and more time devoted to studying mathematics and science is an assumption that most jobs require significant abilities in these fields.

At least with respect to science, however, there is controversy about these claims. Morris Shamos, a professor emeritus of physics at New York University, argues that "widespread scientific literacy is *not* essential to...prepare people for an increasingly technological society" (Shamos, 1988). About 24% of the high school sophomores who are planning to attend college report that they are interested in pursuing a scientific or technical career (Office of Technology Assessment, 1988, Figure 1.1). Shamos does not dispute the need for these students to receive a thorough education in science during their high school years. He argues, though, that there is no need for most citizens and workers to become scientifically literate. A similar argument could be made regarding the need for most students to take algebra, geometry, trigonometry, and statistics.

At least with respect to workers in nontechnical occupations, his view might appear at first glance to be supported by the findings cited above that achievement in science has no effects on wage rates, earnings, or the unemployment of young men and women when other competencies are held constant. Further support for the Shamos position would appear to come from a survey of small- and medium-sized employers who are members of the National Federation of Independent Business (NFIB). When asked how frequently the employees most recently hired by their firm needed to "use knowledge gained of chemistry, physics, or biology" in their jobs, 74% reported that such knowledge was never required; only 12% reported that such knowledge was used at least once a week. Asked how frequently new employees had to "use algebra, trigonometry, or calculus," 68% reported that such skills were never required by the job; only 12% reported that they were used at least once a week (Bishop & Griffin, in press).

However, the skills used by entry-level workers at NFIB firms are not decisive evidence about employers' needs for three reasons. First, the low levels of scientific and mathematical competence in the work force may have obliged companies to postpone technological innovations such as statistical process control that require such skills and to simplify the functions that are performed by workers who lack technical training. If better-educated employees were available, entry-level workers might be given greater responsibility and become more productive. Second, the NFIB sample does not tell us what is happening at large firms and in the jobs occupied by long-tenure employees at small firms. The chief executive officers of many large, technologically progressive companies such as Motorola and Xerox are insisting that their factory jobs now require workers who are much better prepared in mathematics and science than ever before. Third, employers may not realize how the knowledge and skills developed in high school science and mathematics classes contribute to productivity in their jobs. Not knowing which employee possesses which academic skill, employers have no way of learning from experience which scientific and mathematical abilities are helpful in doing a particular job. Science and mathematics are believed to teach thinking, reasoning, and learning skills applicable outside the classroom and the laboratory. If these abilities are developed successfully by these courses, productivity might benefit even when there are no visible connections between job tasks and course content.

The lack of wage and earnings' responses to scientific and mathematical knowledge is also not decisive evidence in favor of Shamos's position; research indicates that differences in worker productivity do not result in proportional differences in wage rates. When people hired for the same or very similar jobs are compared, someone who is 20% more productive than average during the first weeks on a job receives only a 1.6% higher starting wage. After a year's employment, better producers received only a 4% higher wage at nonunion firms with about 20 employees; they had no wage advantage at unionized establishments with more than 100 employees or at nonunion establishments with more than 400 employees (Bishop, 1987).

Employers have good reasons for not varying the wage rates of their employees in proportion to their perceived job performance. All feasible measures of individual productivity are unreliable and unstable. In most cases, only subjective measures of job performance are available. Workers are averse to risk and reluctant to accept jobs in which the judgment of one supervisor can result in a large wage decline in the second year of employment (Hashimoto & Yu, 1980; Stiglitz, 1974). Most productivity differentials are specific to a firm; this reduces the risk that not paying particularly productive workers a comparably higher salary will result in their going elsewhere (Bishop, 1987). Also, pay that is highly contingent on individual performance can weaken cooperation and generate incentives to sabotage others (Lazear, 1989). Finally, in unionized settings, the union's opposition to merit pay will often be decisive.

If relative wage rates only partially compensate the most capable workers for their greater productivity, why do they not obtain promotions or switch to better-paying firms? To some degree, they do, particularly in managerial, professional, technical, and craft occupations. However, because worker productivity cannot be measured accurately and cannot be communicated reliably to other employers, this sorting process is slow and only partially effective. Consequently, when men and women under the age of 30 are studied, the effects of specific skills on wage rates may not correspond to their true effects on productivity; therefore, direct evidence of the impact of specific skills on productivity is required before conclusions can be drawn. We turn, therefore, to direct evidence on the effects of scientific, mathematical, and technical competencies on job performance.

The Impact of Scientific, Mathematical, and Technical Competencies on Worker Productivity

Over the last 50 years, industrial psychologists have conducted hundreds of studies, involving hundreds of thousands of workers, on the relationship between productivity in particular jobs and various predictors of that productivity. They have found that scores on tests measuring competence in reading, arithmetic, and mechanical comprehension are strongly related to productivity in almost all of the civilian jobs studied (Ghiselli, 1973; Hunter, 1983). However, published studies of productivity in civilian jobs generally have not examined the impact of competence in high-school-level science and mathematics (algebra, geometry, and trigonometry) on job performance. On the other hand, for military jobs, there is a great deal of evidence on the effect of competence in high-school-level science and mathematics on job performance.

The Armed Services Vocational Aptitude Battery (ASVAB) contains both a Mathematics Knowledge subtest evaluating understanding of algebra and geometry and a General Science subtest. The focus of the General Science subtest is on facts and concepts; there is minimal coverage of higher-level, scientific problemsolving. The ASVAB also contains four subtests that evaluate an individual's technological competence – mechanical comprehension, electronics information, auto information, and shop information (sample questions are provided in Appendix B).[5] Research using the ASVAB has found that scientific, technical, and mathematical reasoning competencies all have large effects on success in training and on paper-and-pencil measures of job knowledge (Hunter, Crosson, & Friedman, 1985).

But because both the criterion – training success – and the predictor – competence in particular areas – are measured by paper-and-pencil tests, there is a danger that results may be biased by common methods. Therefore, it would be desirable to check these findings in a data set in which ASVAB subtest scores predict a hands-on measure of job performance. Maier and Grafton's (1981) study of ASVAB's (version 6/7) ability to predict the hands-on results of Skill Qualification Test (SQT) provides such a data set. Maier and Grafton described the hands-on SQTs that they used in their study as follows:

> SQTs are designed to assess performance of critical job tasks. They are criterion-referenced in the sense that test content is based explicitly on job requirements and the meaning of the test scores is established by expert judgment prior to administration of the test rather than on the basis of score distributions obtained from administration. The content of SQTs is a carefully selected sample from the domain of critical tasks in a specialty. Tasks are selected because they are especially critical, such as a particular weapon system, or because there is a known training deficiency. The focus on training deficiencies means that relatively few on the job can perform the

> tasks, and the pass rate for these tasks therefore is expected to be low. Since only critical tasks in a specialty are included in SQTs, and then only the more difficult tasks tend to be selected for testing, a reasonable inference is that performance on the SQTs should be a useful indicator of proficiency on the entire domain of critical tasks in the specialty (1981, pp. 4–5).

Table 1 shows the impact of academic and technical competencies on job performance in several Military Occupational Specialties (MOS). (The statistical measure is the standardized regression coefficients in a regression in which ASVAB subtests predict the SQTs.) The data are taken from Appendixes A and B in Maier and Grafton (1981). The regressions were estimated for seven major categories of Military Occupational Specialties – Skilled Technical, Skilled Electronic, General Maintenance and Construction, Mechanical Maintenance, Missile Battery Operators and Food Service Workers, Unskilled Electronic, and Clerical – all of which have close civilian counterparts. (A complete description of the statistical model and results is available in Bishop, 1988b.)

The results for the academic subtests – science, mathematical knowledge, and technical information – are quite different from the wage-rate regression for young males. Scientific knowledge which had small negative effects on wage rates, now has positive effects on hands-on measures of job performance in all seven MOS categories, significantly so in three of them and in the pooled data. With the sole exception of the mechanical maintenance category, the two mathematical subtests – mathematical knowledge and arithmetic problem-solving – have much more impact on SQTs than do scientific knowledge or computational speed. The Mathematics Knowledge subtest, which assesses algebra and geometry, is responsible for most of this. The results of the four "technical" subtests – mechanical comprehension, auto information, shop information, and electronics information – had no impact on job performance in clerical jobs. However, they had very substantial impacts on job performance in all of the other occupations.

Table 2 shows the consequences of a one-population standard deviation (SD) increase in scientific knowledge, mathematical knowledge, arithmetical reasoning, technical information, and word knowledge on worker productivity in standard deviation units. The proportionate change in productivity that results from a one-SD increase in scientific knowledge is probably between 25 and 40% of the numbers given in the table.[6] If we assume that the SD of true productivity averages 30% of the mean wage in these jobs, the average impact of one SD of scientific competency is about 2.0% of the wage. Averaging across all seven occupations, the impact of a simultaneous, one SD increase in both mathematical reasoning subtests is about 6.4%, assuming that the standard

Table 1
Effects of Competencies on Hands-On Measures of Job Performance (Standardized Regression Coefficients)

Military Occupational Speciality	General Science	Math. Know.	Arithmetic Reasoning	Word Knowl.	Comput. Speed	Clerical Checking	Electr. Info.	Shop Info.	Auto Info.	Mechanical Comp.	R^1	N
Skilled Technical	5.7*	12.1***	6.2**	21.5***	3.1	2.4	17.4***	13.2***	1.7	9.2***	0.55	1,324
Skilled Electronic	7.2	26.1***	-2.1	-0.4	-1.3	8.4*	4.5	24.6***	9.8	8.6	0.43	349
General Maintenance (Construction)	13.4***	44.1***	-10.1***	6.6*	6.8**	4.3*	12.1***	11.7***	8.2**	-0.4	0.59	879
Mechanical Maintenance	9.6	6.1	-6.3	-0.4	23.5***	5.5	-8.9	20.6*	31.4***	4.2	0.41	131
Missle Battery Operators & Food Service Workers	7.6*	10.6**	11.4***	6.1	-3.7	5.0	10.0**	6.2	17.9***	10.9**	0.41	814
Unskilled Electronic	-2.5	1.8	5.8*	-1.0	5.3*	3.6	7.7**	6.2*	2.7	0.4	0.05	2,545
Clerical	6.4	20.6***	24.1***	11.8***	8.5**	1.5	6.5	-3.0	8.7***	-6.8	0.43	830

Notes:

* Significant at the 0.1 level using a two-tailed test.
** Significant at the 0.05 level using a two-tailed test.
*** Significant at the 0.01 level using a two-tailed test.
1. A proportion of the variance as explained by the model.
From Bishop (1988b). Reanalysis of Maier and Grafton's (1981) data on the ability of ASVAB (6/7 version) to predict skill qualification test scores. Maier and Grafton corrected their correlation matrix for restriction of range so the coefficients measure the effect of a population, standard deviation change in the test score on job performance in standard deviation units multiplied by 100.

Table 2
Effects of Competencies on Productivity as Percent of the Wage

	Science	Mathematical Reasoning	Technical	Word Knowledge
Nonclerical jobs	2.0	5.3	11.5	1.6
Clerical jobs	1.1	13.4	1.6	3.5
All noncombat jobs	2.0	6.4	10.1	1.9

Note(s):
Unweighted averages of standardized regression coefficients for Table 1 multiplied by 0.3.

deviation for true productivity is 30% of the wage. Averaging over the six nonclerical occupations, a one-SD increase in all four of the technical subtests raises productivity by about 11.5% of the average wage.

These results provide strong support for the claim in *A Nation at Risk* that improved mathematics education for the mass of high school students will improve the productivity of the work force. Doing a better job of teaching arithmetical problemsolving, algebra, and geometry–the mathematical reasoning skills measured by the ASVAB subtests–clearly promises to be worthwhile.

With respect to science, however, the findings are more equivocal. The productivity increase of 2% per SD on the science subtest appears to be modest. This is probably due to the inadequacies of the ASVAB's 11-minute, 24-item General Science subtest. Scientific competence is poorly measured by this short subtest and this causes the estimates to understate the effects of science on job performance. Nevertheless, the data clearly do not disprove Shamos's claim that, for most workers, "widespread scientific literacy is *not* essential to...prepare people for an increasingly technological society" (Shamos, 1988). However, there is a need for new research to determine whether broader and more reliable measures of scientific knowledge and understanding will have more substantial effects on evaluating job performance in nontechnical jobs than is measured by the ASVAB's General Science subtest.

Probably the most striking of the empirical findings is the very large effect of the results of the ASVAB's technical subtests on performance in nonclerical jobs and the wage rates and earnings of young males. These results imply that broad *technical literacy* is essential for workers who operate and/or maintain equipment that is similar in complexity to that used by the armed forces.

Policy Implications

The key to motivating students to learn science, mathematics, and technology lies in recognizing and rewarding learning effort and achievement.

Some students are attracted to the serious study of science, mathematics, and technology by an intrinsic fascination with the subjects. However, they pay a heavy price in the scorn of their peers and lost leisure time. Society offers them little reward for their efforts. But most students are not motivated to study by a love of the subject. Most "don't like to do any more school work than [they] have to" (Public Opinion Laboratory, 1988, Q. AA37N). As a result, far too few high school students put serious time and energy into learning science, mathematics, and technology, and society suffers. If this situation is to be turned around, the peer pressure against studying must be greatly reduced and the rewards for learning must be increased substantially.

The full diversity of types and levels of accomplishment needs to be communicated so that everyone – no matter how advanced or behind – faces a reward for greater time and energy devoted to learning. Learning accomplishments need to be described on an absolute scale so that improvements in the quality and rigor of the teaching and greater effort by all students make everybody better off. Increasing numbers of employers need workers who are competent in science, mathematics, and technology. If these employers know who is well educated in these fields, they will provide the rewards needed to motivate study. Ninety-two percent of 10th graders say, "I often think about what type of job I will be doing after I finish school" (Public Opinion Laboratory, 1988, Q. AA13C). If the labor market were to begin rewarding learning in school, high school students would respond by studying harder, teachers would find teaching more rewarding, parents would demand higher standards, and local voters would be willing to pay higher taxes in order to have better local schools.

Some persons might respond to this strategy for achieving excellence by stating a preference for intrinsic over extrinsic motivation of learning. This, however, is a false dichotomy. Nowhere else in our society do we expect people to devote thousands of hours to a difficult task while receiving *only* intrinsic rewards. Public recognition of achievement and the symbolic and material rewards received by achievers are important generators of intrinsic motivation. They are, in fact, among the central ways a culture symbolically transmits and promotes its values.

The policy recommendations that follow are grouped into five categories:

- Revision of the science, mathematics, and technology curricula
- School-based rewards for learning
- Better indicators of accomplishment in science, mathematics, and technology
- Reform of college admission criteria
- School-sponsored communication of academic achievement to employers.

Revision of Science, Mathematics, and Technology Curricula

Analysis of the causes of American apathy toward the teaching and learning of science, mathematics, and technology has important implications for the curricula. Many of the weaknesses of mathematics and science curricula—the constant review and repetition of old material, the slow pace and minimal expectations—are adaptations to the low level of effort most students are willing to devote to these subjects. When considering proposed revisions of the curricula, the motivation of students to take demanding courses and to study hard must be a central concern.

Another constraint that must be dealt with is the great diversity of the learning goals and capabilities of high school students. On the National Assessment of Educational Progress (NAEP) mathematics scale, 15% of 13-year-olds have better mathematics skills than the average 17-year-old student and 7% of 13-year-olds score below the average 9-year-old (NAEP, 1988a). It is neither feasible nor desirable for all high school students to pursue the same science, mathematics, and technology curricula. Some students will want to pursue natural science in greater depth and rigor than others. Some students will want to concentrate on technology, not pure science. Some courses will be easier than others and, inevitably, students will be able to choose the less demanding courses.

State requirements that students take more mathematics and science courses to graduate will have little effect on learning if students can meet the requirements by taking less exacting courses. Holding background characteristics and the rigor of the mathematics and science courses constant, an *additional* three courses in mathematics and science during high school years increases the gain in mathematics competency between the 10th and 12th grades by only 0.19 of a grade level equivalent and *reduces* science gains by 0.09 of a grade level equivalent and also reduces English and social studies gains by 0.17 to 0.18 of a grade level equivalent. Holding background characteristics and the number of courses constant, taking five college preparatory mathematics and science courses—chemistry, physics, algebra II, trigonometry, and calculus—increases the gain on mathematics and science tests by 0.75 of a grade level equivalent and increased the gain in English and social studies by 0.34 to 0.44 of a grade level equivalent (see note 2 and Bishop, 1985). These data clearly imply that learning rates are determined by the rigor—not the number—of courses taken in a subject.

A further strategy that is bound to fail is that of setting minimum standards for graduation. Because some students arrive in high school far behind others and the consequences of not getting a diploma are so severe, minimum competency standards are never set very high, (and cannot in good conscience be set very high given the constraints on the system). Consequently, few students are challenged by the minimum competency test. Once they satisfy the minimum, many students stop putting effort into their courses.

How, then, do we convince students to work hard in science and mathematics courses? How do we encourage them to select courses that require a lot of work, just to be average achievers? The answers are by *(1) developing rigorous courses that teach students concepts and material that they will use after leaving high school, (2) convincing students that the science and mathematics they are being taught are useful by presenting them as solutions to practical, real-world problems, (3) defining accomplishment in such a way that students who work hard will perceive themselves as successful, and then (4) recognizing and rewarding accomplishment.*

Usefulness is an indispensable criterion for selecting the topics to be included in the science curriculum for these reasons. First, the social benefits of learning derive from the use of the knowledge and skills, not from the fact that they are in one's repertoire. Second, skills and knowledge that are not used deteriorate very rapidly. In one set of studies, students tested two years after taking a course had forgotten one half of the college psychology and zoology, one third of the high school chemistry, and three fourths of the college botany that they had learned (Pressey & Robinson, 1944). Skills and knowledge that are used are remembered. Consequently, if learning is to produce long-term benefits, the competencies developed must continue to be used after the final examination (either in college, the labor market, or elsewhere). Furthermore, usefulness is essential because students do not put energy into learning things that they perceive to be useless. Finally, the labor market will not, in the long run, reward skills and competencies that have no use. Indeed, in most circumstances selecting workers on the basis of competencies that are not useful in a company's jobs is a violation of Title VII of the Civil Rights Act.

Differentiating the Senior High School Curriculum. By the 10th grade most students have a good idea of what types of jobs they want after finishing their education. Ninety-seven percent can select a particular occupation at which they expect to be employed at age 40 and 77% agree with the statement: "I am quite certain about what kinds of jobs I would enjoy doing when I am older" (Public Opinion Laboratory, 1988, Q. AA13C & AA22A). Students who are planning scientific careers need to be able to take college preparatory biology, chemistry, and physics courses that prepare them for the core courses they will face in college. Students not planning scientific careers, however, often fail to see how these courses will be useful to them. When asked to rate, "How useful do you think the [science course you are now taking] will be to you in your career?" on a 5-point scale, 23% of the high school sophomores selected the "No Use" extreme end of the scale; only 28% selected "Very Useful," at the other end of the scale (Public Opinion Laboratory, 1988, Q. AASCI1F).

One approach to this problem is to point out to students how the material in standard college preparatory science courses is useful in nonscientific jobs and in everyday life. Presumably, science teachers try to do this already. Another approach is to modify the standard curricula. That is the approach of the new mathematics and science curricula proposed by the National Council of

Teachers of Mathematics (1989) and the American Association for the Advancement of Science (1989). This makes sense for the first 10 years of schooling. However, there is no standard science curriculum in the 11th and 12th grades, and it is not realistic to propose that everyone take the same courses. At these grade levels, the most effective way to motivate students to take demanding science courses and to study hard is to tailor courses to students' career interests.

Teaching Science and Mathematics by Infusing Them into Technology Courses. The analyses of the labor market success of young men and of job performance in the military presented in Sections 2 and 3 indicate that young people who expect to have jobs in which they use or maintain complicated pieces of equipment should receive a thorough technological education. Computer classes are examples of the kinds of courses needed. High school sophomores described their computer classes as "Very Useful" for their desired career 53% of the time and as of "No Use" only 6% of the time (Public Opinion Laboratory, 1988, Q. AACOMF).

The Principles of Technology (PT) course developed by a consortium of vocational education agencies in 47 states and provinces in association with the Agency for Instructional Technology and the Center for Occupational Research and Development is another example of a course that meets this need very well. This two-year, applied physics course is both academically rigorous and practical. Each six-day subunit deals with the unit's major technical principle (e.g., resistance) as it applies to one of the four energy systems – mechanical (both rotational and linear), fluid, electrical, and thermal. A subunit usually consists of two days of lectures and discussions, a mathematics skills laboratory, two days of hands-on physics application laboratories, and a subunit review. This approach appears to be quite effective for teaching basic physics. In one study comparing regular high school physics students to those enrolled in the Principles of Technology course, the PT students' knowledge of basic physics concepts was substantially below that of the regular physics students on the pretest taken at the beginning of the course. At the end of their senior year, however, the PT students scored an average of 81, compared to an average of 66 for the regular physics students (Perry, 1989). Another study by John Roper (1989) obtained similar results.

School-Based Rewards for Learning

Cooperative Learning. One effective way of persuading peers to value learning and to support effort in school is to reward the group for the individual learning of its members. This is the approach taken in cooperative learning. Research results (Slavin, 1983) suggest that the two key ingredients for successful cooperative learning are:

- A cooperative incentive structure – awards based on group performance – which seems to be essential for students working in groups to become truly involved in tutoring and encouraging each other to study.
- A system of individual accountability, in which everyone's maximum effort must be essential to the group's success and the effort and performance of each group member must be clearly visible to his or her teammates.

For example, students might be grouped into evenly matched teams of four or five members who are heterogeneous in ability. After the teacher presents new material, the team studies together on worksheets to prepare each other for periodic quizzes. The team's score is an average of the scores of the team members; high team scores are recognized on a class bulletin board or through group certificates of achievement.

What seems to happen in cooperative learning is that the team develops an identity of its own and group norms arise that are different from the norms that hold sway in the students' other classes. The group's identity arises from the extensive personal interaction among group members in working toward a shared goal. Because the group is small and the interaction intense, the effort and success of each team member are known to the other teammates. Such knowledge allows the group to reward each team member for his or her contribution to the team goal, and this is what seems to happen.

Laboratory assignments provide a natural environment for cooperative learning, but there is a tendency for one member of the laboratory partnership to do all of the planning and interpreting of the experiment. Somehow, this must be prevented because it results in the slower student not learning from the exercise. (The Slavin [1983] model avoids this problem by assessing students' learning individually and then averaging scores to arrive at the team score.) One approach would be to form teams from constituent laboratory partnerships and to recognize teams on the basis of quizzes given at the end of the laboratory or the following day.

Honoring Academic Achievement. Schools should strengthen their awards and honors systems for accomplishments in science, mathematics, and technology. The medals, trophies, and school letters awarded in interscholastic athletics are powerful motivators for achievement on the playing field. Academic pursuits need a similar system of reinforcement. Awards and honors plans should be designed so that almost every student can receive at least one award or honor before graduation if he or she makes an effort. Outstanding academic performance (e.g., high grades or high test scores) should not be the only standards for defining excellence. Awards could be given for progress made in science and mathematics, for participation in science and mathematics contests, for perfect attendance records in science and mathematics classes, and for student of the week (the criteria could vary weekly). The guideline for making an award should be the criterion referenced: if greater numbers achieve the standard of excellence, more awards should be given. Periodically, the parents of the most recent award winners and sponsoring teachers could be invited to an evening assembly at which time the principal would present the students with a certificate or plaque recognizing their accomplishments.

A prominent place in the school should be reserved for bulletin boards where pictures of the most recent winners and the reasons for their receiving recognition could be posted. Another form of recognition could be displays of student work: science projects, applied technology projects, and so forth.

A National Network of Science, Mathematics, and Technology Clubs. At present, only 3.2% of high school sophomores are members of a science club, only 2.5% are members of a mathematics club and only 1.6% belong to a computer club (Public Opinion Laboratory, 1988, Q. BA10K–BA10M). Memberships in these clubs should be increased and the clubs should be joined together in a national network. The national student organizations should sponsor interschool competitions, visits to science museums, and science and technology project competitions that feed into national competitions such as the Westinghouse Science projects awards. The national organizations would function in much the same way as the national offices of Boy and Girl Scouts, Future Farmers of America, and Vocational-Industrial Clubs of America. They would provide training to teachers and student leaders as well as develop program activity packets to help local science and mathematics teachers devise activities for their clubs.

Better Indicators of Accomplishment in Science, Mathematics, and Technology

Developing Better Assessment Mechanisms. As student recognition and rewards come increasingly to depend on the results of school assessments of competency, it becomes ever more pressing that we improve our methods of assessing mathematical, scientific, and technological achievement. Linking assessment to curriculum also implies a need for a greater diversity of assessment mechanisms. States should not be discouraged from having their own unique curricula simply because examinations keyed to them are not available. However, the need for multiple versions and for fairness to minorities make test development very expensive. The federal government should underwrite state consortia and other organizations that seek to develop alternatives to currently available tests and assessment mechanisms. Priority needs to be given to developing methods of assessing higher-order thinking skills and hands-on performance through simulations, judged portfolios, and demonstrations of skills conducted in front of judges. High cost has been the primary barrier to the use of these richer forms of assessment. Consequently, consideration should be given to subsidizing these more costly assessment mechanisms.

Instituting Statewide Achievement Examinations. States should adopt statewide tests and assessments of competency and knowledge that are specific to the curriculum being taught (e.g., New York State's Regents Examinations) and then give students a competency profile/portfolio certifying their performance on these examinations which could be used as an aid in searching for jobs. Admission to state universities and merit-based scholarships should be based on these achievement examinations and on Advanced Placement tests. In addition to their incentive effects, these examinations/assessments would:

- Inform students and parents better about how well students are doing and thus help parents work with teachers to improve their children's performance.
- Make the relationship between teachers and students more cooperative, so that teachers and students work together to prepare the students for their examination.
- Create a database that school boards and parents could use to evaluate the quality of education being provided by their local schools.
- Enable employers to use scores on these examinations to help improve their selection of new employees. If the uncertainties involved in hiring are reduced, expanding employment will become more profitable, total employment will increase, and recent high school graduates will be able to compete better with more experienced workers for top quality jobs.

Reform College Admission Policies

Expand Advanced Placement Courses. The Advanced Placement (AP) program is a cooperative educational endeavor that offers course descriptions, examinations, and sets of curricular materials in 28 different academic subjects. Students who take these courses and pass the examinations receive college credit for high school work.

The students of Jaime Escalante's Advanced Placement calculus classes have demonstrated how young people from disadvantaged backgrounds can use the AP program as an upward mobility escalator. The student body of the James A. Garfield High School in Los Angeles is composed of predominantly disadvantaged minorities, yet it accounts for 17% of all Mexican Americans taking the AP calculus examination and 32% of all Mexican Americans who pass the advanced version of the test (Berlin & Sum, 1988). There is no secret how they did it; they worked extremely hard. Students signed a contract committing themselves to extra homework and extra time at school—and they lived up to their commitment. What this experience establishes is that minority youngsters from disadvantaged backgrounds can be persuaded to study just as hard as academic track students in Japan, Finland, and England and that, if they do, they will achieve at the same level. Escalante cast aside the zero-sum competition of grades and rank in class and set for his students a very difficult externally defined goal. He convinced them that they could succeed and that there was great honor in taking on the challenge. The success at Garfield High School is replicable.

Expanding the AP program should be a centerpiece of any effort to promote excellence in science, mathematics, and technology education. It clearly meets a need because it is growing rapidly. The numbers of students taking AP tests more than doubled between 1983 and 1988. Nevertheless, only 8,022 of the 22,902 high schools in the United States participate in the Advanced Placement program and only 52 AP tests are taken on average in each participating high school. In the class of 1988, only 2.5% took the AP calculus examination, 0.7% took the AP computer science examination, 1.1% took AP biology, 0.7% took AP chemistry, and 0.6% took AP physics (The College Board, 1988). The nation should set a goal of doubling these percentages every two years for the next decade. New AP tests should be established in principles of technology, electronics, algebra, geometry, trigonometry, probability, statistics, psychology, and business mathematics so that larger numbers of 10th and 11th graders and students planning to attend two-year technical colleges may participate in the AP program. Acting in concert, the college presidents of a large group of selective two-year and four-year colleges should send a letter to every high school principal in the country (with copies to the local school board and local newspaper) urging them to establish AP courses in science, mathematics, and technology. They also should announce that, starting in 1993, students seeking admission to their college should have taken and passed at least one AP course in their junior year and be taking more than one AP course in their senior year.[7]

The federal government can facilitate the growth of the AP program by underwriting the development of AP examinations for new subjects, by financing summer institutes for the teachers of AP courses, and by offering a $100 AP excellence award (larger if the student is eligible for Pell Grant aid) to every student who gets a 3 or above on an AP test and a $150 award for getting a top score. To ensure that attending a summer institute is considered desirable, the compensation should be generous. In 1988 approximately 42,000 teachers taught these courses. Rapid expansion of the program will require a yearly increase of 20,000 in the stock of teachers teaching AP courses. If 30% of the increment to the stock were to experience summer institute training for 6 weeks, the cost would be about $42 million. In 1988, 286,009 students would have been eligible for an AP excellence award, so the program would have cost under $40 million. If widespread publicity were attached to these awards, they would prompt a major expansion of the program.

Dropping SAT and Substituting AP and State Scores. In the future, it would be preferable for all concerned if colleges and universities could be encouraged to drop from their selection criteria the results of the Scholastic Aptitude Test (SAT) and to substitute the scores attained on AP and state-sponsored achievement examinations and assessments, which reflect the curricula taught in high school.

While national tests are necessary, the SAT is not the kind of test that is most helpful. It suffers from two very serious limitations: the limited range of the achievements that are evaluated and its multiple-choice format. The test was designed to be curriculum-free. To the extent that it evaluates students' understanding of material taught in school, the material it covers is vocabulary and elementary and junior high school mathematics. Most of the college preparatory subjects studied in high school – science, social science, history, geography, literature, technology, art, music, computers, calculus, probability, and statistics – are completely absent from the test. As a result, the SAT fails to generate incentives to take the more demanding courses or to study hard. The test advertises itself as an ability test, but, in fact, is an achievement test measuring a very limited range of achievements (Jencks & Crouse, 1982). Jencks and Crouse have recommended that either the SAT evaluate a much broader range of achievements or it be dropped in favor of an expanded set of AP examinations. Knowledge and understanding of literature, history, technology, science, and higher-level mathematics should all be assessed. These examinations should not be limited to a multiple-choice format; essays and extended answers should be required where appropriate.

School-Sponsored Communication of Academic Achievement to Employers

At present, recent high school graduates are not considered for jobs with good wages and promotion opportunities because applicants who lack extensive work experience are not considered for such positions. Extensive work experience is considered essential partly because it contributes to productivity, but also because it provides indications of competence and reliability that employers use to identify who is most qualified. Recent high school graduates have no such record. Information on a student's high school performance is generally not available, so the entire graduating class appears to American employers as one undifferentiated mass of unskilled and undisciplined workers. New York Life Insurance Company had such difficulty finding qualified clerical employees in the United States that the company moved some of its claims processing work to Ireland. A supervisor at New York Life Insurance Company commented on television, "When kids come out of high school, they think the world owes them a living" (Public Broadcasting System, 1989). Surely this generalization does not apply to every graduate, but the students who are disciplined and academically well prepared currently have no way of communicating this fact to prospective employers. The result is that 29% of the June 1986 white high school graduates not attending college did not have a job the following October. Minority youth must overcome even more damaging stereotypes and they generally lack a network of adult contacts who can provide leads and references for jobs. The result is that 58% of the black 1986 high school graduates not attending college did not have a job in October 1986.

If employers had access to more information on young people's accomplishments in high school, graduates would not be relegated to sales clerk jobs simply because of their age. Like their peers in Ireland, Europe, Canada, and Japan, they should be allowed to compete for attractive jobs on the basis of the knowledge and skills they have gained in high school. Employers should start requiring high school transcripts and giving academic achievement (particularly achievement in mathematics and science) much greater weight in hiring. Business and industry should communicate this policy to schools, parents, and students.

Releasing Student Records. Schools can help students get good jobs by developing an equitable and efficient policy for releasing their records. School officials have the dual responsibility of protecting students' rights to privacy and of helping them to find good, suitable employment. Students and their parents should receive copies of transcripts (encased in plastic) and other important records so that they can make them available to anyone they choose. Schools also could develop guidelines that explain their rights to parents and students as well as the advantages and disadvantages of disclosing information.

According to the Family Educational Rights and Privacy Act of 1974, all that a student/graduate needs to do to have school records sent to a prospective employer is to sign a form specifying the purpose of disclosure, which records are to be released, and who is to receive the records. The waiver and record request forms used by employers contain this information so, when such a request is received, the school is obliged to respond. Requiring that graduates fill out a school-devised form—as did one high school I visited—results in employers' not getting the transcript requested and graduates' not getting the job. There are probably millions of high school graduates who do not realize that they failed to get a job they were hoping for because their high school did not send the transcript. Schools can serve students best by handling all inquiries expeditiously and without charge.

Credential Databank and Employee Locator Service. It may be unrealistic to expect 22,902 high schools to develop efficient systems of maintaining student records and responding quickly to requests for transcripts. An alternative approach would be to centralize the recordkeeping and dissemination functions in a trusted, third-party organization. This organization would be easy to regulate and, thus, everyone could be assured that privacy mandates were being observed. The student would determine which competencies to have assessed and what types of information to include in his or her competency portfolio. Competency assessments would be offered for a variety of scientific, mathematical, and technological subjects, languages, writing, business economics, and occupational skills. Tests with many alternative forms (or computer-administered, based on a large item bank) would be used so that students could retake the test a month later, if desired. Only the highest score would remain in the system. Students would be encouraged to include descriptions of their extracurricular activities, their jobs, and any other accomplishments they feel are relevant and to submit samples of their work such as research papers, art work, or pictures of a project made in metal shop. Files could be updated after leaving high school.

Students would have three different ways of transmitting their competency profile to potential employers. First, they would receive certified copies of their portfolio which they could take to interviews or mail to prospective employers. Second, they would be able to call an 800 number and request that their portfolio be sent to specific employers. Third, they could ask to be placed in an employee-locator databank similar to the student-locator services operated by the Educational Testing Service and American College Testing Program. A student seeking a summer or post-graduation job would specify the type of work sought and dates of availability. Employers seeking workers could ask for a print-out of the portfolios of all the individuals living near a particular establishment who have expressed interest in that type of job and who pass the employer's competency screens. Student-locator services have been used extensively by colleges trying to recruit minority students and an employee-locator service could be used in the same way. This would increase the rewards for diligent study significantly because an employee-locator service is likely to result in a bidding war for all qualified students – especially minorities – whose portfolios are in the system.

Summary

The keys to motivating students to learn science, mathematics, and technology are recognizing and rewarding learning effort and achievement. Learning accomplishments need to be described on an absolute scale so that improvements in the quality and rigor of the teaching and greater effort by all students help everybody. We should establish individual learning goals that challenge students to the maximum extent possible. Achievement of these goals should be recognized at school awards ceremonies and communicated to college admissions officers and the labor market. If employers know who is well educated in science, mathematics, and technology, they will provide the rewards necessary to motivate learning. There are tens of thousands of teachers and hundreds of thousands of students willing to work as hard as Jaime Escalante and his students. To release this energy, excellence must be defined by an external standard, small teams of students and teachers must be challenged to achieve excellence, and they must be honored and rewarded when they succeed. If we create such a system, the positive results will be astonishing.

Notes

1. The American College Testing Program examination is a marginal improvement over the Scholastic Aptitude Test in this respect because two of the four tests are on "Social Studies Reading" and "Natural Science

Reading." However, most of the questions in these two subtests are based on reading passages contained in the test booklet. Of the 52 questions on each of the subtests, only 15 are to be answered "on the basis of your previous schoolwork in the subject."

2. A study was conducted of the cohort of High School and Beyond students projected to graduate in 1982. The dependent variables were the changes in test scores and grades between sophomore and senior years. The model included extensive controls for variables that may influence both curricula and outcomes. Holding the total number of academic courses and their distribution across fields constant, taking the five college preparatory mathematics and science courses – chemistry, physics, algebra II, trigonometry, and calculus – raised mathematics, science, verbal, and social studies tests score gains, but *lowered* students' grade point average significantly (Bishop, 1985).
3. The National Longitudinal Survey data set is not the only data set in which the wage rates of young adults are shown to be little affected by academic competencies. Similar results were obtained in Willis and Rosen's (1979) analysis of the earnings of those who chose not to attend college, in the National Bureau of Economic Research-Thorndike data, Kang and Bishop's (1986) analysis of High School and Beyond seniors, and Bishop, Blakemore, and Low's (1985) analysis of both the class of 1972 and High School and Beyond data. Bishop, Blakemore, and Low studied the effect of mathematics, reading, and vocabulary test scores on the wage rates and earnings of high school graduates for both 1972 and 1980 in a model that contained controls for grade point average and the number of credit hours of academic and vocational courses. In both these years, none of the variables representing academic performance – the three test scores, the grade point average and the number of academic courses – had a significant (at the 10% level) effect on the wage rate of the first, post-high school job. Only one variable (the vocabulary test for female members of the class of 1972) had a significant effect on the wage rate 18 months after graduation.
4. The survey was of a stratified random sample of the membership of the National Federation of Independent Business. Larger firms had a significantly higher probability of being selected for the study. The response rate to the mail survey was 20% and the number of usable responses was 2,014 (Bishop & Griffin, in press).
5. Even though the Armed Services Vocational Aptitude Battery was developed as an "aptitude" test, the current view of testing professionals is that "achievement and aptitude tests are not fundamentally different....Tests at one end of the aptitude-achievement continuum can be distinguished from tests at the other end primarily in terms of purpose. For example, a test for mechanical aptitude would be included in a battery of tests for selecting

among applicants for pilot training because knowledge of mechanical principles has been found to be related to success in flying. A similar test would be given at the end of a course in mechanics as an achievement test intended to measure what was learned in the course" (Wigdor & Gardner, 1982).

6. Studies that measure output for different workers in the same job at the same firm, using physical output as a criterion, can be manipulated to produce estimates of the standard deviation of nontransitory output variation across individuals. This standard deviation averages about 0.14 in operative jobs, 0.28 in craft jobs, 0.34 in technician jobs, 0.164 in routine clerical jobs, and 0.278 in clerical jobs with decisionmaking responsibilities (Hunter, Schmidt, & Judiesch, 1988; Bishop, 1988a). Because there are fixed costs to employing an individual (facilities, equipment, light, heat, and overhead functions such as hiring and payroll), the coefficient of variation of marginal products of individuals is assumed to be 1.5 times the coefficient of variation of productivity. Because about two thirds of clerical jobs can be classified as routine, the coefficient of variation of marginal productivity for clerical jobs is 30% [1.5*(.33*.278+.67*.164)]. Averaging operative jobs with craft and technical jobs produces a similar 30% figure for blue collar jobs. The details of and rationale for these calculations are explained in Bishop, 1988a and 1988b.

7. This proposal sounds radical but, in fact, it is only a modest change from current practice at prestigious colleges. A survey of college placement officials conducted by *USA Today* and my own interviews at Cornell University and the State University of New York at Binghamton found that students were expected to take Advanced Placement (AP) courses if they are offered and grade point averages were adjusted for the level of difficulty of the courses. Students and parents are often not aware of this policy, however, and many have not factored it into their course selections. The announcement would have two effects: it informs students and parents of existing admissions policies and warns them that, in 1993, those seeking admission to the most desirable colleges will be held accountable even if a local high school does not offer AP courses. This announcement will generate strong political pressure on principals and school boards to expand their AP program and to allow additional students to take AP courses. Students at schools not offering AP courses should be offered other ways of demonstrating college-level proficiency, such as an AP independent study option, taking courses during the summer at a local college, or high scores on achievement examinations such as the New York State Regents examinations, California's Golden State examinations, or the SAT examinations. Exceptions would have to be made for students from underrepresented minorities, foreign students, and other individual cases, but exceptions should not become the rule.

References

Altonji, J. (1988). *The effects of high school curriculum on education and labor market outcomes.* Evanston, IL: Northwestern University, Center for Urban Affairs and Policy Research.

American Association for the Advancement of Science. (1989). *Science for all Americans: A Project 2061 report on literacy goals in science, mathematics, and technology.* Washington, DC: Author.

Berlin, G. & Sum, A. (1988). *Toward a more perfect union: Basic skills, poor families, and our economic future* (Occasional Paper No. 3). New York: Ford Foundation Project on Social Welfare and the American Future.

Bishop, J. (1985). *Preparing youth for employment.* Columbus: Ohio State University, National Center for Research in Vocational Education.

Bishop, J. (1987). The recognition and reward of employee performance. *Journal of Labor Economics, 5*(4, Pt. 2), S36–S56.

Bishop, J. (1988a). *The economics of employment testing* (Working Paper No. 88-14). Ithaca, NY: Cornell University, Center for Advanced Human Resources Studies.

Bishop, J. (1988b). *The productivity consequences of what is learned in high school* (Working Paper No. 88-18). Ithaca, NY: Cornell University, Center for Advanced Human Resources Studies. (In press, *Journal of Curriculum Studies.*)

Bishop, J., Blakemore, A., & Low, S. (1985). *High school graduates in the labor market: A comparison of the class of 1972 and 1980.* Columbus: Ohio State University, National Center for Research in Vocational Education.

Bishop, J., & Griffin, K. (in press). *Recruitment, training and skills of small business employees.* Washington, DC: American Express Company.

The College Board. (1988). *AP Yearbook 1988.* New York: Author.

Frederick, W. C. (1977). The use of classroom time in high schools above or below the median reading score. *Urban Education, 11*(4), 459–464.

Frederick, W., Walberg, H., & Rasher, S. (1979). Time, teacher comments, and achievement in urban high schools. *Journal of Educational Research, 73*(2), 63–65.

Friedman, T., & Williams, E. B. (1982). Current use of tests for employment. In A. K. Wigdor & W. R. Gardner (Eds.), *Ability testing: Uses, consequences, and controversies, part II: Documentation section* (pp. 99–169) (Report of the National Research Council Committee on Ability Testing). Washington, DC: National Academy Press.

Ghiselli, E. E. (1973). The validity of aptitude tests in personnel selection. *Personnel Psychology, 26*, 461–477.

Goodlad, J. (1983). *A place called school.* New York: McGraw-Hill.

Hashimoto, J., & Yu, B. (1980). Specific capital, employment and wage rigidity. *Bell Journal of Economics, 11*(2), 536–549.

Hunter, J. E. (1983). *Test validation for 12,000 jobs: An application of job classification and validity generalization analysis to the general aptitude test battery.* Washington, DC: U.S. Department of Labor Employment Service.

Hunter, J. E., Crosson, J. J., & Friedman, D. H. (1985). *The validity of the Armed Services Vocational Aptitude Battery (ASVAB) for civilian and military job performance.* Washington, DC: U.S. Department of Defense.

Hunter, J. E., Schmidt, F. L., & Judiesch, M. K. (1988). *Individual differences in output as a function of job complexity.* Iowa City: University of Iowa, Department of Industrial Relations and Human Resources.

International Association for the Evaluation of Education Achievement. (1988). *Science achievement in seventeen nations.* New York: Pergamon Press.

Jencks, C., & Crouse, J. (1982). Aptitude vs. achievement: Should we replace the SAT? *The Public Interest,* pp. 21–35.

Kang, S., & Bishop, J. (1986, Spring). The effect of curriculum on labor market success immediately after high school. *Journal of Industrial Teacher Education, 94(1),* 15–29.

Lazear, E. P. (1989, June). Pay equality and industrial politics. *Journal of Industrial Teacher Education, 97*(3), 561–580.

Maier, M. H., & Grafton, F. C. (1981). *Aptitude composites for ASVAB 8, 9 and 10.* (Research Rep. 1308). Alexandria, VA: U.S. Army Research Institute for Behavioral and Social Sciences.

McKnight, C.C., Crosswhite, F. J., Dossey, J. A., Kifer, E., Swafford, J. O., Travers, K. J., & Cooney, T. J. (1987). *The underachieving curriculum: Assessing U.S. school mathematics from an international perspective.* A National Report on the Second International Mathematics Study. Champaign, IL: Stipes Publishing Co.

Meyer, R. (1988). *The effects of high school academic reforms on course enrollments and academic skills growth: Evaluation and new directions for reform* (National Assessment of Vocational Education Technical Report). Madison: University of Wisconsin, La Follette Institute of Public Affairs.

National Assessment of Educational Progress. (1988a). *The mathematics report card.* Princeton, NJ: Educational Testing Service.

National Assessment of Educational Progress. (1988b). *The science report card.* Princeton, NJ: Educational Testing Service.

National Commission on Excellence in Education. (1983). *A nation at risk: The imperative for educational reform.* Washington, DC: U.S. Government Printing Office.

National Council of Teachers of Mathematics. (1989). *Curriculum and evaluation standards for school mathematics.* Reston, VA: Author.

National Opinion Research Corporation. (1982). *High school and beyond survey.* Chicago, IL: Author.

Nielsen Company, A. C. (1987). [Time spent watching television by children and teens]. Unpublished raw data.

Office of Technology Assessment. (1988). *Elementary and secondary education for science and engineering.* Washington, DC: U.S. Government Printing Office.

Organization for Economic Cooperation and Development. (1986). *Living conditions in OECD countries: A compendium of social indicators* (Social Policy Studies No. 3). Paris, France: Author.

Perry, N. J. (1989, June). The new improved vocational school. *Fortune,* pp. 127–138.

Pressey, S., & Robinson, F. (1944). *Psychology and New Education.* New York: Harper.

Public Opinion Laboratory. (1988). *Longitudinal survey of American youth: Base year codebook (Fall 1987-Spring 1988).* Dekalb, IL: Author.

Quinlan, J. (Producer), & Merrow, J. (Director). (1989). *Learning in America.* [Television]. United States: WETAVISION.

Roper, J. (1989, January). Technology creates a new physics student. *The Physics Teacher,* 27(1), 26–28.

Rosenbaum, J., & Kariya, T. (1987, August). *Market and institutional mechanisms for the high school to work transition in Japan and the U.S.* Paper presented at the meeting of the American Sociological Association, Chicago, IL.

Shamos, M. (1988, November 23). The flawed rationale of calls for "literacy." *Education Week,* p. 28, 22.

Sizer, T.R. (1984). *Horace's compromise: The dilemma of the American high school.* Boston: Houghton Mifflin.

Slavin, R. (1983). When does cooperative learning increase student achievement? *Psychological Bulletin, 99(3)*, 429–445.

Stevenson, H. (1983). *Making the grade: School achievement in Japan, Taiwan and the United States* (Annual report). Stanford, CA: Center for Advanced Study in the Behavioral Sciences.

Stevenson, H., Lee, S., & Stigler, J. W. (1986). Mathematics achievement of Chinese, Japanese and American children. *Science, 231*, 693–699.

Stiglitz, J. E. (1974). Risk sharing and incentives in sharecropping. *Review of Economic Studies, 61*(2), 219–256.

U.S. Military Entrance Processing Command. (1984). *Counselor's manual for the armed services vocational aptitude battery, form 14* (DoD 1304.12X). Chicago, IL: U.S. Department of Defense.

Wigdor, A. K., & Gardner, W. R. (Eds.). (1982). *Ability testing: Uses, consequences, and controversies, part I.* (Report of the National Research Council Committee on Ability Testing). Washington, DC: National Academy Press.

Willis, R., & Rosen, S. (1979, October). Education and self-selection. *Journal of Political Economy, 87*, S7–S36.

APPENDIX A

The scientific, mathematical, and technical knowledge of American high school students is generally recognized to be very low. The National Assessment of Educational Progress (NAEP) reports that only 7.5% of 17-year-old students can "integrate specialized scientific information" (NAEP, 1988b, p.51) and that only 6.4% "demonstrated the capacity to apply mathematical operations in a variety of problem settings" (NAEP, 1988a, p. 42).

There is a large gap between the science and mathematics competence of young Americans and their counterparts overseas. In the 1960s, the low ranking of American high school students in such comparisons was attributed to the fact that the test was administered to a larger proportion of American than European and Japanese youth. This is no longer the case. In the Second International Mathematics Study, the universe from which the American sample was drawn consisted of high school seniors taking a college preparatory mathematics course. This group represented 13% of the age cohort, a proportion that is roughly comparable to the 12% of Japanese youth who were in their sample frame and is considerably smaller than the 19% of youth in the Canadian province of Ontario and the 50% of Hungarians who took the test. In algebra, the mean score for this very select group of American students was about equal to the mean score of the much larger group of Hungarians and substantially below the Canadian achievement level (McKnight, et al 1987).

The findings of the Second International Science Study are even more dismal. Table A-1 compares the scores in the types of courses in biology, chemistry, and physics against the proportion of the 18-year-old population to which the international test was administered. For example, the 25% of Canadian 18-year-olds taking chemistry know just as much chemistry as the very select 1% of American high school seniors taking their second chemistry course (most of whom are in Advanced Placement courses). The 28% of Canadians taking biology know much more than the 6% of American 18-year-olds who are taking their second biology course (International Association for the Evaluation of Educational Achievement, 1988).

The poor performance of American students is sometimes blamed on the nation's "diversity." Many affluent parents apparently believe that their children are learning acceptably by international standards. This is not the case. In Stevenson, Lee, and Stigler's (1986) study of fifth grade mathematics achievement, the best students in the 20 classrooms sampled in Minneapolis were outstripped by the students in every single classroom studied in Sendai, Japan and by students in 19 of the 20 classrooms studied in Taipeh, Taiwan. The nation's top high school students rank far behind much less elite samples of students in other countries. At the end of secondary school, the gaps in scientific knowledge between white American students and their counterparts in England, Canada, Finland, and Japan are as large as the gap between black and white students in the United States. In mathematics, the gaps are more than twice as large as the gap between black and white students in this country. The learning deficit is pervasive.

Table A-1
Comparison of Performance on Science Tests of 18-Year-Olds in the United States and Other Countries (In U.S. Standard Deviation Units) 1983

	Percent Taking Biology	Biology Score	Percent Taking Chem.	Chem. Score	Percent Taking Physics	Physics Score
Australia	18	0.67	12	0.49	11	0.19
Canada	28	0.52	25	-0.04	19	-0.37
England	4	1.66	5	1.74	6	0.81
Finland	45	0.91	14	-0.24	14	-0.48
Hong Kong (F6)*	7	0.84	14	1.46	14	0.87
Hungary	3	1.42	1	0.55	4	0.70
Italy	14	0.29	2	0.02	19	-1.11
Japan	12	0.54	16	0.78	11	0.67
Norway	10	1.10	15	0.23	24	0.46
Poland	9	1.24	9	0.38	9	0.38
Singapore	6	1.88	5	1.55	7	0.59
Sweden	15	0.69	15	0.13	15	-0.04
United States	6	–	1	–	1	–

Notes:
The scores reported in the table were calculated by first subtracting the U.S. mean score for students taking their second year of the subject from the scores for other nations and then dividing by the standard deviation of scores in the U.S. sample.
* Form 6 or English system which is six years of high school as compared to the Chinese system which is five years.
From Science Achievement in Seventeen Nations, by International Association for the Evaluation of Educational Achievement, 1988, New York: Pergamon Press.

Appendix B

Reproduction of Excerpts from
Counselor's Manual for the Armed Services
Vocational Aptitude Battery

Form 14

COUNSELOR'S MANUAL
FOR THE
ARMED SERVICES VOCATIONAL
APTITUDE BATTERY
FORM 14

JULY 1984

DoD 1304.12X

Purposes

The ASVAB is a multiple aptitude battery designed for use with students in Grades 11 and 12 and in postsecondary schools. The test was developed to yield results that are useful to both schools and the military. Schools use ASVAB test results to provide educational and career counseling for students. The military services use the results to identify students who potentially qualify for entry into the military and for assignment to military occupational training programs.

Like other multiple aptitude batteries, the ASVAB measures developed abilities and predicts what a person could accomplish with training or further education. This test is designed especially to measure potential for occupations that require formal courses of instruction or on-the-job training. In addition, it provides measures of general learning ability that are useful for predicting performance in academic areas.

The ASVAB can be used for both military and civilian career counseling. Scores from this test are valid predictors of success in training programs for enlisted military occupations. Through the use of validity generalization techniques, predictions from military validity studies can be generalized to occupations that span most of the civilian occupational spectrum. Although some enlisted occupations are military specific, more than 80% of these occupations have direct civilian occupational counterparts.

Since the ASVAB was first used in high schools in 1968, it has been the subject of extensive research and has been updated periodically. Appendix A contains a brief history of the ASVAB and the various forms that have been used.

Key Features

ASVAB-14, introduced in the 1984-85 school year, contains several key features that were not included in previous forms. These key features include

- **improved usefulness in measuring vocational aptitudes:** In addition to yielding *academic composites* that provide measures of academic potential, ASVAB-14 supplies *occupational composites* that provide measures of potential for successful performance in four general career areas.
- **increased reliability:** Changes in the length and number of subtests have increased the test's reliability without a substantial increase in testing time.
- **nationally representative norms:** ASVAB-14 is normed on a nationally representative sample of 12,000 women and men, ages 16-23, who took the test in 1980.

Content

Subtests

The ASVAB consists of 10 subtests. Eight are power subtests that allow maximum performance with generous time limits. Two subtests are speeded.

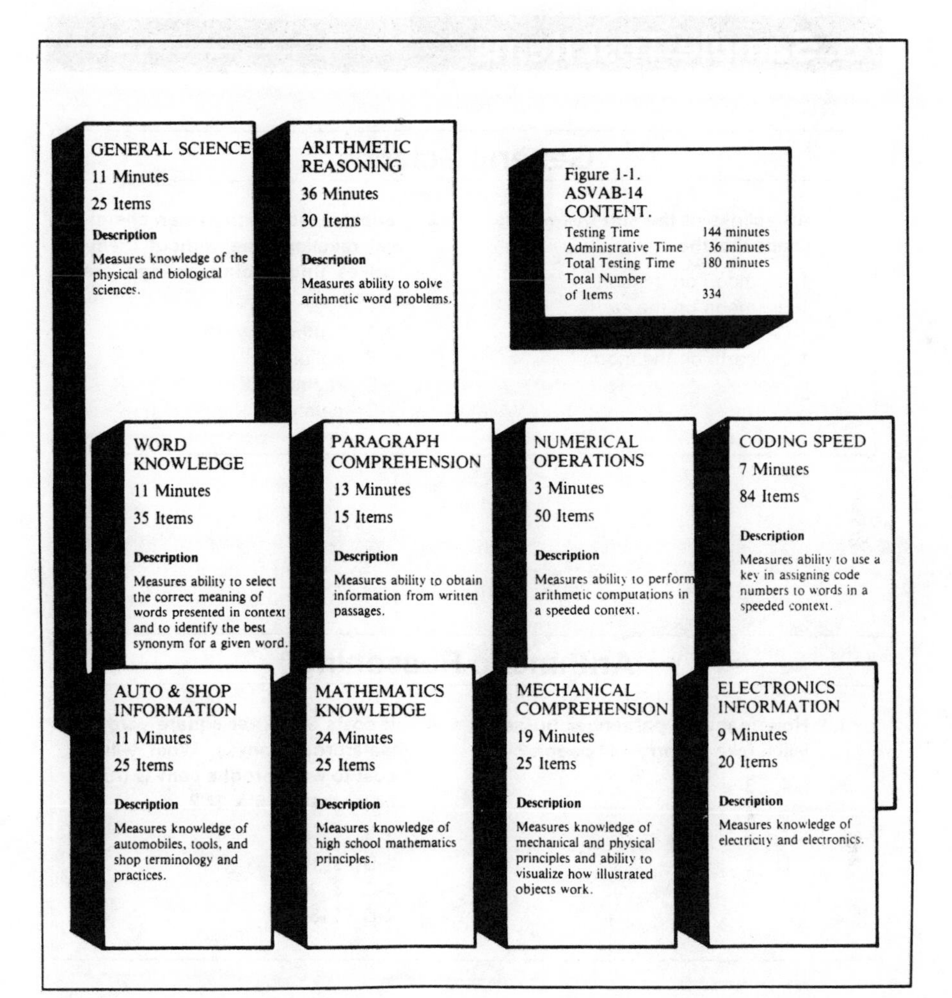

GENERAL SCIENCE
11 Minutes
25 Items
Description
Measures knowledge of the physical and biological sciences.
ARITHMETIC REASONING
36 Minutes
30 Items
Description
Measures ability to solve arithmetic word problems.
WORD KNOWLEDGE
11 Minutes
35 Items
Description
Measures ability to select the correct meaning of words presented in context and to identify the best synonym for a given word.
PARAGRAPH COMPREHENSION
13 Minutes
15 Items
Description
Measures ability to obtain information from written passages.
NUMERICAL OPERATIONS
3 Minutes
50 Items
Description
Measures ability to perform arithmetic computations in a speeded context.
CODING SPEED
7 Minutes
84 Items
Description
Measures ability to use a key in assigning code numbers to words in a speeded context.
AUTO & SHOP INFORMATION
11 Minutes
25 Items
Description
Measures knowledge of automobiles, tools, and shop terminology and practices.
MATHEMATICS KNOWLEDGE
24 Minutes
25 Items
Description
Measures knowledge of high school mathematics principles.
MECHANICAL COMPREHENSION
19 Minutes
25 Items
Description
Measures knowledge of mechanical and physical principles and ability to visualize how illustrated objects work.
ELECTRONICS INFORMATION
9 Minutes
20 Items
Description
Measures knowledge of electricity and electronics.

Figure 1-1. ASVAB-14 CONTENT.

Testing Time	144 minutes
Administrative Time	36 minutes
Total Testing Time	180 minutes
Total Number of Items	334

B. Sample Test Items

General Science

1. **An eclipse of the sun throws the shadow of the**

   1-A moon on the sun.
   1-B moon on the earth.
   1-C earth on the sun.
   1-D earth on the moon.

2. **Substances which hasten chemical reaction time without themselves undergoing change are called**

   2-A buffers.
   2-B colloids.
   2-C reducers.
   2-D catalysts.

Arithmetic Reasoning

3. **How many 36-passenger busses will it take to carry 144 people?**

   3-A 3
   3-B 4
   3-C 5
   3-D 6

4. **It costs $0.50 per square yard to waterproof canvas. What will it cost to waterproof a canvas truck cover that is 15′ x 24′?**

   4-A $ 6.67
   4-B $ 18.00
   4-C $ 20.00
   4-D $180.00

Word Knowledge

5. **The wind is <u>variable</u> today.**

   5-A mild
   5-B steady
   5-C shifting
   5-D chilling

6. **<u>Rudiments</u> most nearly means**

   6-A politics.
   6-B minute details.
   6-C promotion opportunities.
   6-D basic methods and procedures.

Paragraph Comprehension

7. Twenty-five percent of all household burglaries can be attributed to unlocked windows or doors. Crime is the result of opportunity plus desire. To prevent crime, it is each individual's responsibility to

7-A provide the desire.
7-B provide the opportunity.
7-C prevent the desire.
7-D prevent the opportunity.

8. In certain areas water is so scarce that every attempt is made to conserve it. For instance, on one oasis in the Sahara Desert the amount of water necessary for each date palm tree has been carefully determined.

How much water is each tree given?

8-A no water at all
8-B water on alternate days
8-C exactly the amount required
8-D water only if it is healthy

Numerical Operations

9. 3 + 9 =

9-A 3
9-B 6
9-C 12
9-D 13

10. 60 ÷ 15 =

10-A 3
10-B 4
10-C 5
10-D 6

Coding Speed

KEY

bargain 8385	house 2859	owner 6227
chin 8930	knife 7150	point 4703
game 6456	music 1117	sofa 9645
	sunshine 7489	

QUESTIONS	A	B	C	D	E
11. game	6456	7150	8385	8930	9645
12. knife	1117	6456	7150	7489	8385
13. bargain	2859	6227	7489	8385	9645
14. chin	2859	4703	8385	8930	9645
15. house	1117	2859	6227	7150	7489
16. sofa	7150	7489	8385	8930	9645
17. owner	4703	6227	6456	7150	8930

	A	B	C	D	E
18. music	1117	2859	7489	8385	9645
19. knife	6227	6456	7150	7489	8485
20. sunshine	4703	6227	6456	7489	8930
21. chin	1117	2859	4703	7150	8930
22. sofa	4703	6227	7150	8485	9645
23. bargain	2859	6456	8385	8930	9645
24. point	1117	4703	6227	6456	7150

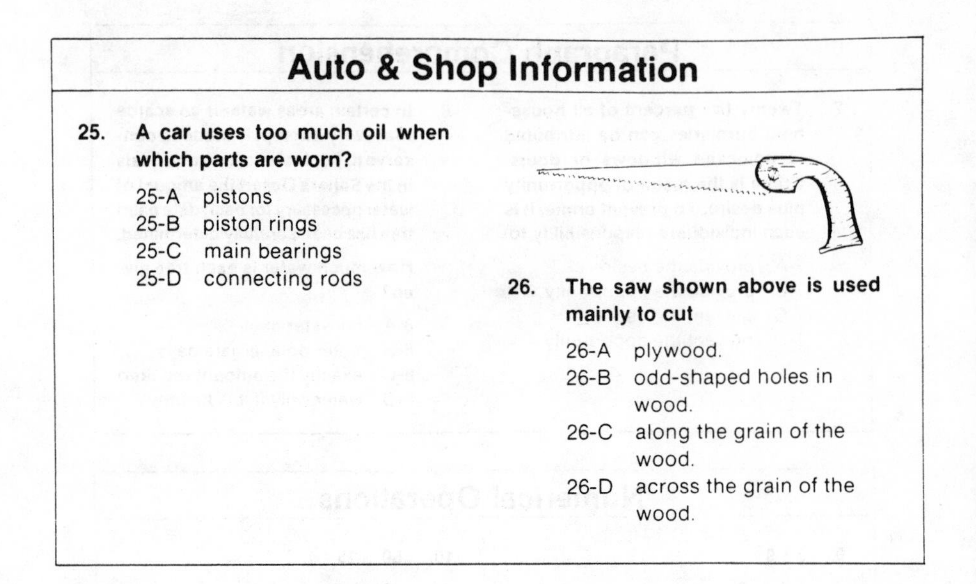

Auto & Shop Information

25. A car uses too much oil when which parts are worn?

25-A pistons
25-B piston rings
25-C main bearings
25-D connecting rods

26. The saw shown above is used mainly to cut

26-A plywood.
26-B odd-shaped holes in wood.
26-C along the grain of the wood.
26-D across the grain of the wood.

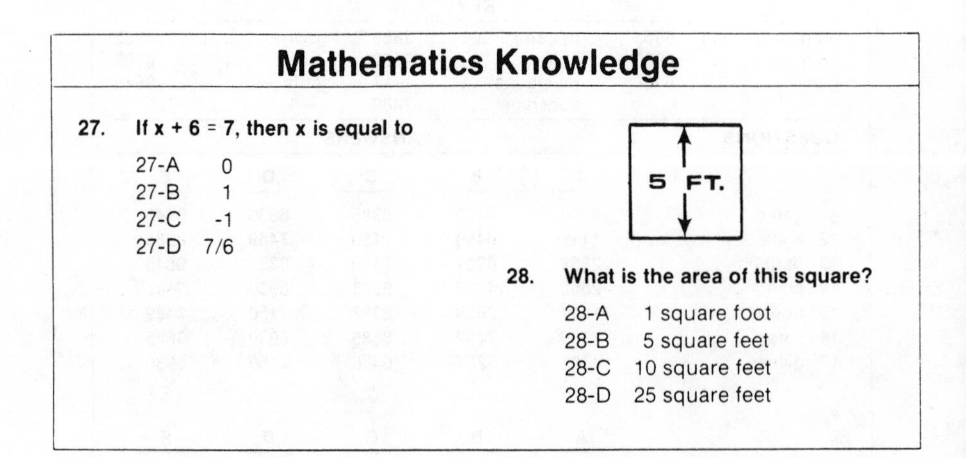

Mathematics Knowledge

27. If x + 6 = 7, then x is equal to

27-A 0
27-B 1
27-C -1
27-D 7/6

28. What is the area of this square?

28-A 1 square foot
28-B 5 square feet
28-C 10 square feet
28-D 25 square feet

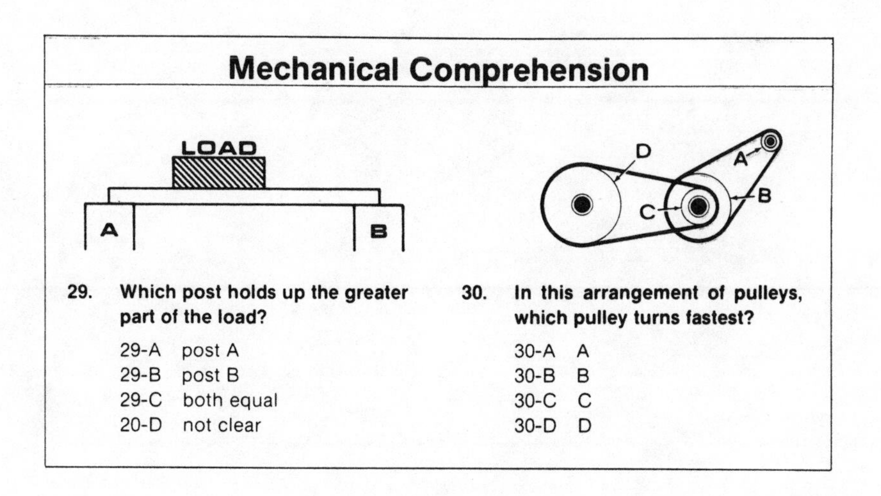

Mechanical Comprehension

29. **Which post holds up the greater part of the load?**

    29-A post A
    29-B post B
    29-C both equal
    20-D not clear

30. **In this arrangement of pulleys, which pulley turns fastest?**

    30-A A
    30-B B
    30-C C
    30-D D

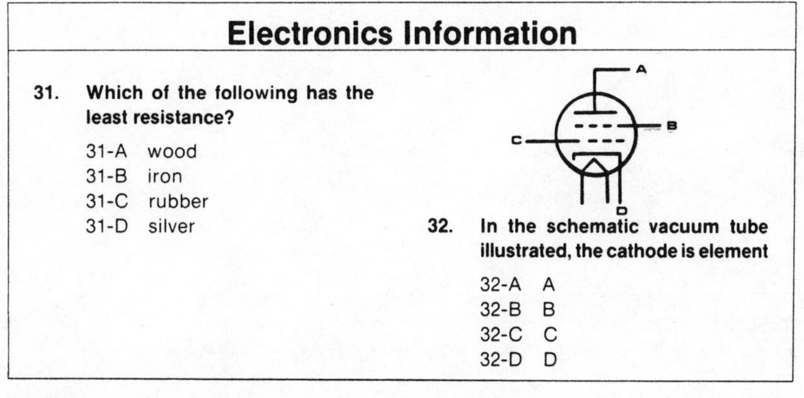

Electronics Information

31. **Which of the following has the least resistance?**

    31-A wood
    31-B iron
    31-C rubber
    31-D silver

32. **In the schematic vacuum tube illustrated, the cathode is element**

    32-A A
    32-B B
    32-C C
    32-D D

Key To The Sample Test Items

1.	B	17.	B
2.	D	18.	A
3.	B	19.	C
4.	C	20.	D
5.	C	21.	E
6.	D	22.	E
7.	D	23.	C
8.	C	24.	B
9.	C	25.	B
10.	B	26.	B
11.	A	27.	B
12.	C	28.	D
13.	D	29.	A
14.	D	30.	A
15.	B	31.	D
16.	E	32.	D

4

SCIENTIFIC LITERACY, SOCIETAL CHOICES, AND IDEOLOGIES

Gerard Fourez

This chapter reviews aspects of the ideology – concepts, attitudes, and values – related to "scientific literacy" in recent reports on school science. Each report is summarized briefly, its ideological position examined, and the basis for choices among alternative positions discussed. Finally, the implications of these choices for the individual and for society are considered.

The reports reviewed are:

- The Phase I report of the American Association for the Advancement of Science's Project 2061, which consists of six documents – an integrated report, *Science for All Americans,* and five panel reports: Biological and Health Sciences, Mathematics, Physical and Information Sciences and Engineering, Social and Behavioral Sciences, and Technology – intended to help in the formulation of a vision "of what Americans want their schools to achieve" (American Association for the Advancement of Science [AAAS], 1989, p. 3).
- *Science Objectives: 1990 Assessment (Science Objectives),* published by the National Assessment of Educational Progress (NAEP), which provides the framework for an assessment that will determine the nation's grade in science education on the next "Nation's Report Card."

- *Science and Technology Education for the Elementary Years: Frameworks for Curriculum and Instruction (Science Education for the Elementary Years),* a set of recommendations for elementary school science curricula from the National Center for Improving Science Education (NCISE).
- *Science for Ages 5–16 (Science 5–16),* which presents the proposals of the Secretary of State for Education and Science and the Secretary of State for Wales, United Kingdom, Department of Education and Science and the Welsh Office (DESWO).

Perspective and Methodology

I use the term "ideology" to connote a meaning generally accepted among philosophers which is *broader than the common American understanding of the term* (the latter often has a pejorative overtone). According to the so-called "theory of ideologies," an "ideological discourse" relates to any language that presents a vision of the world and society. It operates – implicitly or explicitly, consciously or not – to motivate people, to create cohesion in a community, and to legitimize some social practices, while being unable to explain all of its presuppositions. Language is ideological when there are two or more ways of looking at a topic, and each view has different societal implications (Winograd & Flores, 1987). Each of these perspectives, without being necessarily more false or true in the relevant context, produces different societal effects. In this sense, as Ricoeur (1974) has shown, *ideological language has a positive function necessary to every society and is always present in scientific discourse* (Ricoeur, 1974; Fourez, 1982, 1985, 1988a, 1988b).

In this review, I will emphasize when choices have been made and when there could have been an alternative position. My intent is not to judge the differences. Clearly, I prefer some to others. Rather, my goal is to show some of the societal choices made in these diverse approaches to "scientific literacy." Obviously, as any epistemologist will testify, such an analysis is never completely neutral. Every person speaks from his or her own position. A universal perspective does not exist. I will try to let the documents speak for themselves and to show their ideological choices. My objective is not to declare that some choices would not be the right ones, but to help the reader see that choices are made and that scientific literacy creates an ideological field.

Much of the same vision is shared by the four documents reviewed, but nuances can be enlightening. After having presented the vision the reports have of scientific literacy, I will examine their particular ideologies on these topics: science, technology, mathematics, society, and historical perspectives. To emphasize the ideological choices, I will exaggerate somewhat the contrasts between the perspectives and the central position of each document – sometimes leaving in the shadows the *correction statements* that are often introduced in the reports for the sake of balancing the global orientation.

Scientific Literacy: A Loose Concept?

As the review of the four documents will indicate, the concept of scientific literacy is a slippery one. Each document has its own idiosyncratic definition of scientific literacy, stated either explicitly or implicitly. The choice of definition has implications both for the way the curriculum is subsequently developed and for achieving the desired outcome with respect to the way students will live their individual and collective lives.

The conception of scientific literacy in *Science Objectives* is contained in the following quotation:

> Science educators generally agree that the primary purpose of school science is to cultivate scientific literacy; however, there is far less agreement as to what constitutes scientific literacy or how such a definition might be used to guide the development of meaningful curriculum and instruction. At the very least, there appears to be a consensus among educators that school science should help students acquire the knowledge, skills, and understandings necessary to fulfill their human, social, and economic responsibilities (NAEP, 1989, p. 8).

Science for All Americans offers this view of the scientifically literate person:

> One who is aware that science, mathematics, and technology are interdependent human enterprises with strengths and limitations; understands key concepts and principles of science; is familiar with the natural world and recognizes both its diversity and unity; and uses scientific knowledge and scientific ways of thinking for individual and social purposes (AAAS, 1989, p. 4).

The view of scientific literacy in *Science Education for the Elementary Years* is captured by the following statement:

> Curriculum and instruction should provide children with appropriate experiences with science and technology that enhance their sensibilities about the natural and technological world, improve their skills of inquiry and problemsolving, develop their understanding and appreciation for the limits and possibilities of science and technology, and contribute to their civility in the conduct of human affairs....civility expresses two important goals. The first relates to being civil in personal interactions; that is, being humane and appreciating differences between individuals and their ideas. The

> second concerns the decisions and responsibilities of citizenship. (NCISE, 1989b, pp. 8–9).

The perspective of *Science 5–16* is slightly different. It is concerned with a science curriculum which includes technology that should (a) enable children, if they wish, to choose a career in which that subject is essential, (b) be helpful in their adult life, for example, in the home, in leisure activities, and in their understanding of and approach to health, and (c) more generally be a basis for coping with a world of rapid scientific and technological changes, and for enriching life.

The four perspectives are similar. Some nuances appear, however. *Science Objectives* emphasizes that pupils have to acquire knowledge, skills, and understandings to fulfill their responsibilities. *Science for All Americans* is centered more on a "scientific culture." Children have to be aware of science, understand concepts, be familiar with the natural world, and use science. Although promoting a similar culture, *Science Education for the Elementary Years* is more student-centered than science- or expert-centered. And, as indicated earlier, *Science 5–16* is mainly a science curriculum.

Behind this consensus and underlying these nuances is a question to ponder: Given this goal of scientific literacy, what will be the benefits to individuals and to society? The four reports assume that personal and civic decisionmaking go hand in hand. Still, there is a difference between helping students "to fulfill their human, social, and economic responsibilities" and helping students to be autonomous. There also could be some significance to the fact that *Science for All Americans* adds "especially the minority children on whom the nation's future is coming to depend" (AAAS, 1989, p. 11). Here, the choice is to mention less privileged minorities – not for their own sake, however, but for the "nation's future."

Science for All Americans explains at length what it means by scientific literacy. It summarizes Phase 1 of Project 2061 (2061 is the year Halley's Comet returns) and concentrates on the "substance of scientific literacy." Its goal is to spell out the "knowledge, skills, and attitudes all students should acquire" (AAAS, 1989, p. 3) by the end of high school. Its basic assumption is that schools do not need to teach more content, but to focus on scientific literacy. As quoted earlier, this view is:

> based on the belief that the scientifically literate person is one who is aware that science, mathematics, and technology are interdependent human enterprises with strengths and limitations; understands key concepts and principles of science; is familiar with the natural world and recognizes both its diversity and unity; and uses scientific knowledge and scientific ways of thinking for individual and social purposes (AAAS, 1989, p. 4).

Whereas *Science Education for the Elementary Years* intends "to demystify science by redefining the scientific and technological knowledge in terms of major concepts that are accessible to the majority of people..." (NCISE, 1989b, p. 24), *Science for All Americans* recommends that the traditional boundaries between scientific disciplines be softened and their connections emphasized; that ideas and thinking skills be given more importance at the expense of specialized vocabulary. It recommends the inclusion of topics less commonly found in school curricula, such as the mutual relationship of science, mathematics, and technology as well as their relationship with the social system in general. It also calls "for some knowledge of the most important episodes in the history of science and technology, and of the major conceptual themes that run through almost all scientific thinking" (AAAS, 1989, p. 5).

In its final section, *Science for All Americans* recommends: (a) reducing the amount of material covered in order to weaken rigid subject-matter boundaries, connecting science, mathematics, and technology, and presenting the scientific endeavor as a social and human enterprise; (b) actively engaging students by starting "with questions about phenomena rather than with answers to be learned," by "placing a premium on students' curiosity and creativity," and by establishing positive conditions for change with adequate public support for a decade or longer; and (c) involving all of the social factors connected with teaching in a collaborative reform of education.

While arguing for scientific literacy, *Science for All Americans* vacillates between two orientations. One is centered on self-fulfillment and the other on socioeconomic needs. It states, on the one hand, that the highest purposes of education are: to prepare people to lead personally fulfilling and responsible lives, to become compassionate human beings, to be able to think for themselves and face life head-on, and to participate in societal life. On the other hand, it states that America's future depends on the quality of education. These two orientations are placed in a global perspective and issues ranging from population growth to pollution, including extreme inequities in the distribution of wealth, are considered. However, no specific analysis of the interactions between these two orientations is provided nor is it suggested that sometimes they could be in conflict.

In *Science for All Americans,* scientific literacy is seen as a necessary condition for a future beneficial to humankind. For *Science for All Americans,* science, scientific habits of mind, technological principles, knowledge, intelligent respect for nature, critical and independent thinking, a sound basis for assessing the use of new technologies, and tools for dealing with problems are necessary to provide that which is needed for a promising outlook for a better world. In view of such a need, the report claims that "most Americans are not scientifically literate" (AAAS, 1989, p. 13). Among the causes of national illiteracy are: schools that are "stuck in the nineteenth century" (AAAS, 1989, p. 14), elementary school teachers whose education in science and mathematics is inadequate, communities that do not provide adequate support to science and mathematics teachers, science textbooks that do not encourage active learning and thus impede progress toward scientific literacy, and overstuffed and undernourished science curricula that fail to keep track of what aspects of science, mathematics, and technology it is truly essential to know. *Science for All Americans* thus attempts "to establish a conceptual base for reform by defining the knowledge, skills, and attitudes all students should acquire as a consequence of their total school experience..." (AAAS, 1989, p. 14).

The Nature of Science: How Many Natures Does Science Have?

Each report advocates that all children understand the nature of science. But each report focuses on a different aspect of understanding, thus reflecting past and present debates between epistemologists and philosophers of science.

On one side, *Science for All Americans* and *Science Objectives* propose a view of science that is centered around the notion of "scientific process," or "methodology." They also reflect what could be called an empiricist vision of science.[1] On the other side, *Science Education for the Elementary Years* centers its approach around the concept of "explanation" (theory), whereas *Science 5–16* does not define science by its methods, but by the historical scientific endeavor.

Science 5–16 gives the most historical definition of the "nature of science":

> Pupils should develop an understanding that science is a human activity, that scientific ideas change through time, and that the nature of scientific ideas and the uses to which they are put are affected by the social and cultural contexts in which they are developed (DESWO, 1988, p. vi).

Science Objectives, by contrast, emphasizes the "nature of scientific processes": observing, classifying, inferring, interpreting data, formulating hypotheses, and designing experiments. In *Science Objectives,* emphasis is given to observation, whereas *Science 5–16* does not mention observation in its section on the nature of science; instead, its speaks of ideas and theories related to a specific historical context.

Behind these discourses, one can guess at the existence of two kinds of philosophies of science. In *Science Objectives,* as in empiricist philosophies, it is assumed that from observations, generally conceived of as "objective," scientists build theories. However, *Science 5–16,* probably influenced by Popper and other more recent philosophers, emphasizes that ideas and theories will be changed when they come to be regarded as inadequate in a given *context.*

In this regard, the cover picture of *Science Objectives* is an interesting choice which conveys a probably unintended impression. It represents some laboratory equipment depicted in a modern and stylistic way, thus imparting the perception that science deals with serious things which happen only in a laboratory and are very different from daily life. By contrast, *Science for All Americans* seems to have chosen its pictures to express the point of view that scientific literacy has much to do with human experience and art. Meanwhile, industrial society does not appear very prominently in any of these pictures.

An analysis of the nature of science is carried out at length in *Science for All Americans.* As in *Science Objectives,* the philosophy is *objectivist* and *empiricist*: "through the use of the intellect, and with the aid of instruments that extend the senses, people can discover patterns in all of nature" (AAAS, 1989, p. 25). The basic assumption relates to the paradigm of physics as it has dominated ideologically the philosophy of science (Prigogine & Stengers, 1988); indeed, the universe is viewed as "a vast single system in which the basic rules are everywhere the same" (AAAS, 1989, p. 25). In a similar vein, "fundamentally, the various scientific disciplines are alike" practically in all of their methods, but "scientists differ greatly from one another..." (AAAS, 1989, p. 26). Science is a process that depends "both on making careful observations of phenomena and on inventing theories for making sense out of those observations" (AAAS, 1989, p. 26). Therefore, in an accumulative process, "increasingly accurate approximations can be made to account for the world and how it works" (AAAS, 1989, p. 26). It is assumed, however, that "science cannot provide complete answers to all questions" (AAAS, 1989, p. 26).

Observation is presented mainly as a passive process of trying to see *what happens*, even when scientists actively probe the world and control the conditions of observation. When speaking of observation, *Science for All Americans* does not acknowledge the many studies that emphasize how observation is always an *interpretation* which depends on accepted theories, contexts, and ideas (Austin, 1962; Habermas, 1979; Searle 1979; Quine, 1981, Winograd & Flores, 1987) to give it meaning. Everything in *Science for All Americans* is assumed to conform to the principles of logical reasoning. Scientists have to invent hypotheses or theories to imagine how the world works and this is "as creative as writing poetry, composing music, or designing skyscrapers" (AAAS, 1989, p. 27). Finally, "the essence of science is validation by observation" (AAAS, 1989, p. 28). Interpretation seems to be a source of biases that scientists try to avoid. It is never seen as the foundation of every kind of observation.

The scientific enterprise is represented as nonauthoritarian, as "in the long run...theories are judged by their results" (i.e., to explain more phenomena or to answer "more important questions") (AAAS, 1989, p. 28). Nothing is said about the criteria by which it is decided what questions are more important or who decides what is important; judging by the results, there seems to be an obvious and neutral process. Some will see characteristics of a *technocratic point of view.* Indeed, a perspective is called "technocratic" when it assumes that scientific or technological results can decide societal issues without the mediation of any human and political negotiation. Generally, technocrats believe that science provides *neutral knowledge and rationality* that advance the solution of problems.

In more than one respect, *Science for All Americans,* as well as *Science Objectives,* sound technocratic.

Science is a complex social activity. As such, "science inevitably reflects social values and viewpoints....Scientists are employed by universities, hospitals, business and industry, government, independent research organizations, and scientific associations" (AAAS, 1989, p. 29). However, no mention is made at this point in the document of the military as an employer of scientists.

Science for All Americans puts forward several ethical principles generally accepted in the conduct of science. Some deal with the ethical traditions of scientists in their practice (like accuracy, openness, etc.). Others consider the possible harm that could result from scientific experiments. Still others are concerned with "the possible harmful effects of applying the *results* [italics added] of research" (AAAS, 1989, p. 30). Even though these results "may be unpredictable,...some idea of what *applications* [italics added] are expected from scientific work can be ascertained by knowing who is interested in funding it" (AAAS, 1989, p. 30). (Note the categories of scientific *results,* followed by *applications*.) Only when dealing with applications does *Science for All Americans* mention that science can be connected to the military.

Even though *Science for All Americans* contains some analysis of science as a societal process, the portrayal of science in that report could be the basis for a technocratic ideology. Science is based on neutral observation of a well-organized, lawful universe and science gives results that can be applied in utilitarian ways. For the most part, the influence of society on science remains external.

Science Education for the Elementary Years elaborated its concept of science from another point of view. In its pursuit of scientific literacy, its enemy is the notion of the nature of science in which "science is primarily presented as a body of knowledge and only secondarily as a process for establishing new knowledge" (NCISE, 1989b, p. 13). Moreover, *Science Education for the Elementary Years* rejects the view of science as "disciplinary, cumulative, and largely independent of the processes used to develop new scientific knowledge" (NCISE, 1989b, p. 13). Such an approach leads *Science Education for the Elementary Years* to state that "science originates in *questions* [italics added] about the natural world" (NCISE, 1989b, p. 13). Here, questions precede *observations,* as the draft of the report expressed well: "Neither science nor technology is subject to a singular method, such as that often presented in textbooks as 'the scientific method'....Observation and experimentation are examples, as are processes unique to the domain of technology, such as consideration of cost and risk in technological solutions" (NCISE, 1989a, p. 3.5).[2]

What about mathematics and science? *Science Education for the Elementary Years, Science Objectives,* and *Science 5–16* all present an approach to science in which mathematics is scarcely integrated. According to *Science for All Americans,* mathematics can be valued for its beauty and intellectual challenge, or for its application. Mathematics is universal in a sense that other fields of human thought are not. The ideological function in our society of presenting mathematics as universal should be analyzed (Bloor, 1976). While science is presented in a social context in *Science for All Americans,* the section on mathematics does not mention the social production of that discipline. It would be difficult, for example, to realize from *Science for All Americans* that the military is financing fundamental studies in mathematics with a view to improving the ability to communicate and for secret coding and decoding. Another perspective on mathematical literacy could have analyzed how the language of mathematics, with its notions of rigor, seriousness, precision, and organization can parallel the discourse of economists when they speak of social management.

The ways in which I have analyzed the four reports could be challenged. But, from the analysis I have provided, we can build a table (Table 1) which contrasts the language used by people when they speak of the nature of science and of technology. These contrasts could be used to test textbooks or teachers' discussions.

The Nature of Technology: Does Science Differ from Technology?

For *Science for All Americans,* "In today's world, technology is a complex social enterprise that includes not only research, design, and crafts but also finance, manufacturing, management, labor, marketing, and maintenance....But the results of changing the world are often complicated and unpredictable....Anticipating the effects of technology is therefore as important as advancing its capabilities" (AAAS, 1989, p. 39).

If, in the past, technology grew out of personal experience, now, in the view of *Science for All Americans,* it draws on science and contributes to it. On the same page, *Science for All Americans* presents two contrasting definitions of "engineering": (a) it is "the systematic application of scientific knowledge in developing and applying technology" and (b) it "consists of construing a problem and designing a solution for it" (AAAS, 1989, p. 40). Each definition relates to a specific current in the philosophy of technology, the first being more "science-oriented" (engineering as "applied science"), the second "problem-centered" (according to this view, science can be viewed as an intellectual technology).

After a refined analysis, *Science for All Americans* concludes: "Scientists seek to show that theories fit the data; mathematicians seek to show the logical proof of abstract connections; engineers seek to demonstrate that designs work" (AAAS, 1989, p. 40). Engineering decisions affect the social system more directly than scientific research and they "involve social and personal values as well as scientific judgments" (AAAS, 1989, p. 40). Finally, the philosophy of

Table 1
Contrasting Ideologies of Science and Technology

Science begins with observations.	Science begins with questions.
Science starts with an observation, nonbiased, if possible.	Science originates in contextual ideas and theories.
Science is disciplinary, cumulative, and results-centered.	Science originates in contextual questions.
Science is defined by method.	Science is a historical human process and unique event.
The world is an organized universe with one basic set of rules.	We construct theories and ideas to organize the world from our context.
The universe has one universal rationality.	Rationality is related to human points of view and projects.
There is only one scientific method.	There is a diversity of methods related to contexts and objectives.
Science and technology are clearly distinct.	Science can be seen as an intellectual technology.
Technologies resemble tools that can be used or not.	Technologies are social as well as material systems.
Technology is applied science.	Technology is an intellectual knowledge system by itself.
Scientific theories are *discovered* whereas technologies are *invented.*	People invent scientific models as well as technologies.

Science for All Americans tends to carry the common ideology that presents scientific practice as neutral while technology is value laden. By contrast to *Science for All Americans'* position, the statement could be made that scientists seek to invent theoretical models that will work in the contexts in which they are meant to be used.

Science for All Americans takes the position that the essence of engineering is design under constraint. Some constraints are presented as absolute (like physical laws); whereas others are more flexible (like economic, political, social, and ethical ones). Technological creativity always implies a compromise between conflicting points of view. "The task is to arrive at a design that reasonably balances the many trade-offs" (AAAS, 1989, p. 41). The reader may wonder about the meaning of the word "reasonably." Whose reason or whose criteria will be predominant? One also may wonder about the meaning of *absolute constraints.* After all, the history of science and technology shows how humankind made these constraints "relative" (e.g., there was a time when the absolute constraint of physical laws meant that a heavy body had to fall, but today we have airplanes). Rather than speak of the absolute constraints of physical laws, another ideological orientation could have been conveyed by the statement: "We relate some constraints to limits expressed by scientific laws and principles." But, in this section also, *Science for All Americans* shows its technocratic orientation.

Technologies, according to *Science for All Americans,* always have side effects (which can be positive or negative). They may turn out to be "unacceptable to a substantial fraction of the population, resulting in conflict between groups in the community. To minimize such side effects, planners are turning to systematic risk analysis" (AAAS, 1989, p. 42). This introduces the topic of technology and society and, with it, many ideological choices. For example, *Science for All Americans* emphasizes (free enterprise ideology obliges) that "individual inventiveness is essential to technological innovation. Nonetheless, social and economic forces strongly influence what technologies will be undertaken, paid attention to, invested in, and used" (AAAS, 1989, p. 43). A contrasting ideology could have been expressed as: "Nowadays, social and economic forces are essential to technological innovation, even if some individual inventiveness remains important."

Science for All Americans observes that, because of the economic value of technology, the social system imposes some restrictions on the openness of science and engineering. Here, finally, some connections between science, technology, national security, and the military are emphasized. Significantly, however, the only ethical issues raised with respect to these interactions relate to secrecy, which goes against the ethos of scientists.

Decisions about the use of technology are, however, complex. Sometimes, "the *use* [italics added] of some technology becomes an issue subject to public debate and possibly formal regulation" (AAAS, 1989, p. 44). It is assumed implicitly in *Science for All Americans* that technology can be conceptualized adequately as a tool that can be used or not used. There is no hint of the ways in which organization and social dynamics are constituent parts of technology; for example, railroad technology is not a tool that can be used or not; it is a way of organizing societal life.

With respect to public participation, *Science for All Americans* states: "Individual citizens may seldom be in a position to ask or demand answers to these questions on a public level, but their knowledge of the relevance and importance of answers increases the attention given to the questions by private enterprise, interest groups, and public officials. Furthermore, individuals may ask the same questions with regard to their own use of technology....The cumulative effect of individual decisions can have as great an impact on the large-scale use of technology as pressure on public decisions can" (AAAS, 1989, p. 45). Here, *Science for All Americans* conveys a political philosophy that reflects the dominant ideology of an individualistic society according to which what matters is the *cumulative effect of individual decisions* rather than *politically organized groups of citizens.* The image of society conveyed is of an aggregate of separate individuals. Another option could have spoken of organized associations.

Finally, *Science for All Americans* states that, in making technological decisions, "political factors are likely to have as much influence as technical ones" (AAAS, 1989, p. 45). However, scientists have the special role of presenting the relevant data necessary to evaluate these decisions and to monitor technological development. Nothing is said, though, about *an education in the social sciences or about the nature of institutions* that would be necessary for scientists to fulfill that special role.

Science Objectives, while assessing science's aims, chooses not to speak of technology (in the same way, it neglects the history of science and the societal or mathematical contexts of science). *Science 5–16* is aware that it is not well prepared to deal with technology education. In its framework, technology is specified as a topic only for children from the ages of 5 to 11. After those ages, it is included in the "technological and social aspects" of science. The complete structure of *Science 5–16* is indeed a *science* curriculum. Only one attainment target for ages 11 to 16 relates science with its context, "Pupils should develop a critical awareness of the ways that science is applied in their own lives and in industry and society, of its personal, social and economic implications, benefits and drawbacks" (DESWO, 1988, p. 68). No mention is made of technology for its own sake. It is seen mainly as "applied science....[even if] [t]echnology, however, is more than applied science—it draws the knowledge it needs for solving problems from many disciplines" (DESWO, 1988, p. 7). But even that statement implies that technology is applied knowledge, originating elsewhere. Furthermore, the dynamics of technology, with their economic, social, and cultural overtones, are not analyzed at any level comparable with the analysis contained in *Science for All Americans.*

For *Science Education for the Elementary Years,* "technology is an important area of study; and the emphasis on technology (within the curriculum) balances the emphasis placed on science" (NCISE, 1989b, p. 9). The draft of *Science Education for the Elementary Years* says: "Unless otherwise stated,...[we use] the term *science* to mean both science and technology" (NCISE, 1989a, p. 1.14). The report regrets that, at present, "when technology is introduced, it is usually defined as the application of scientific knowledge" (NCISE, 1989b, p. 14). It also deplores the fact that "current programs present very little information about the relationships between science and technology....Technology is typically defined early in a text series and seldom mentioned after that" (NCISE, 1989b, p. 15).

For *Science Education for the Elementary Years,* "modern science and technology are inextricably bound..." (NCISE, 1989b, p. 15). "Science proposes explanations for questions about the natural world,...technology proposes solutions for problems of human adaptation to the environment" (NCISE, 1989b, p. 18). Later in the report, however, the distinction between science and technology, explanations and solutions, does not play a significant role. The discussion proceeds as if the statement about science had been: *Science proposes mental representations for human adaptation to the environment.* Such a statement would amount to equating both science and technology as proposing solutions to problems of human adaptation to the environment. Many scientists rebuff the idea that science could be equated with an *intellectual technology.* But, epistemologically, such an approach could be more solid than one that defines science as proposing "explanations for questions about the natural world." After all, questions never occur with respect to the *natural world,* but with respect to *a culturally filled world.* To state it approximately, the notion of the natural world, as we use it today, is an abstraction invented culturally in the 16th and 17th centuries. All of the pedagogical recommendations in *Science Education for the Elementary Years* show that, in effect, it does not operate from a concept of science centered on a so-called "natural" world, but from a concept of science that originates on the *questions of children—and that is a culturally constructed world.* For all practical purposes, even if *Science Education for the Elementary Years* makes a distinction between science and technology, it uses an implicit definition of science as a human process seeking solutions to the adaptation of human beings to their environment. In my opinion, this implicit epistemological position, one that treats science and technology in the same way, is one of the strengths of *Science Education for the Elementary Years.* One can wonder why *Science Education for the Elementary Years* did not produce the theory of its practice. Is there a taboo in our society that forbids the mingling of science and technology as a unified process? If so, for what social reasons?

Thus, essentially, *Science Education for the Elementary Years* provides a framework for considerable integration of science and technology into the curriculum. It does it to a greater extent than *Science for All Americans.* The reason for the difference could be that, for *Science Education for the Elementary Years,* scientific literacy is centered on the situation and the questions of pupils "asking questions and identifying problems are the initial steps in the domain of science and technology" (NCISE, 1989, p. 22), whereas for *Science for All Americans* the focus seems to be a societal expectation about "the knowledge, skills, and attitudes all students should acquire" (AAAS, 1989, p. 3). The concept *scientific* used as a norm also appears in the questionnaire of the brochure of the 1989 AAAS Forum for School Science. The *scientific* character of each item is emphasized. There is no such item as: "The scientifically literate high school graduate is able to construct and use an operational model for action, even if the latter differs from current scientific models."

Human Society: An Aseptic Utopia?

Scientific literacy obviously refers to an image of society at large. Each report refers to society, but with unequal emphasis. For *Science Objectives,* the main societal background of scientific literacy is competition with other nations on the one hand, and the state of the work force in the United States on the other. *Science 5–16* does not place much emphasis on society as such, even if there is an attainment target on "technological and social aspects" of science. However, in the chapter "Science for All," some social inequalities are considered. Special attention is given to certain groups which "often find it difficult to realize their full potential in science. These groups include girls, low attainers, gifted pupils, pupils from different ethnic backgrounds, and those with special educational needs" (DESWO, 1988, p. 91), for example, the physically or psychologically disabled. Socioeconomic privileges or handicaps are not mentioned.

Science for All Americans aims at presenting a complete overview of what coherent scientific literacy would be. It offers "a comprehensible picture of the world that is consistent with the findings of the separate disciplines within the social sciences" (AAAS, 1989, p. 77). The chapter on human society states: "The views presented here are based principally on scientific investigation" (AAAS, 1989, p. 77). I believe this is a good example showing (a) how scientific investigation can be ideologically biased and (b) how an observation always depends on one's theoretical assumptions and social position. Reading that chapter, it is indeed difficult to believe that it describes a human society where there is poverty and hunger (even in the United States). In the aseptic utopia described there as "human society," there is not a word about poverty, hunger, racism, sexism, the third world, military intervention, or the arms race. Does scientific investigation or observation have to forget such phenomena and reject them as unscientific? The conceptual framework used is such that these phenomena, which are reality for many, become invisible. If the children of the most powerful nation of the world become *literate* in this way, I am afraid that the rest of the world, and the poor in America, will suffer. The blindness of that section is even stranger given that, later, when the panel analyzes its own actions in society, it seems quite aware of conflicting social analysis: "What schools can accomplish for many children is very limited as long as a quarter of the students are raised in poverty, drug use and violence go unabated, racism persists, and commercial television remains vapid or worse while educational television stays chronically undernourished.…Only if some of today's worst social problems are ameliorated will the schools be able to take the sweeping reform steps that will enable them to have extensive positive effects on society. The reforming of education and the reforming of society need to go hand in hand" (AAAS, 1989, pp. 157–158).

The content of the chapter on the relationship between science and society reflects the dominant beliefs (ideologies) of the American individualistic and technocratic culture. This was perhaps unavoidable, but one wonders if a person can be considered scientifically literate with respect to human society without having some understanding of what many in this world consider – rightly or wrongly, that is not the debate here – as poverty, class struggle, or any equivalent concept. Likewise, some critical ideas that go beyond stereotypes of socialist ideologies could be helpful for a scientifically literate person of the 20th century. Finally, in a report covering the subject of scientific and technological literacy, I expected more analysis of the ways in which social and technical systems interact.

Science Education for the Elementary Years, which is centered on children and not on topics, does not speak much about society. It does not analyze society's complexity, but situates the goals of science education in the perspective of a social analysis: "The next decade of population expansion in America will see a significant increase in the cultural diversity of our youth population and an alarming increase in the number of children being born into poverty" (NCISE, 1989b, p. 20). Its assumption is that, through the present educational system, "students are aware of pollution, energy, disease, and other issues, and they are willing to make changes in their lives to solve those problems, but they do not feel that their actions could have an impact on the world's problems" (NCISE, 1989b, p. 17). In *Science Education for the Elementary Years,* "decisions" are not there to be handled in a technocratic way, but to be made by people. This report emphasizes that the issues of science, technology, and society in education are still unsolved, which brings forth the difficulty of changing "attitudes and assumptions about teaching science" (NCISE, 1989b, p. 17).

Historical Perspectives: A History of Persons and Events?

The history of science seems virtually absent from *Science Education for the Elementary Years'* definition of scientific literacy. This document presents a completely atemporal view of science – which is a choice.

Science Objectives gives little place to the history of science, even if the panel believes "that students should be cognizant of the major developments in science history" (NAEP, 1989, p. 14). Science assessments should test the "familiarity with *persons and events* [italics added] that have helped shape contemporary scientific understanding in various content areas" (NAEP, 1989, p. 14). Will such a perspective take into account the *socioeconomic-epistemological dynamics* of the construction of science?

For *Science for All Americans,* historical perspectives are an integral part of scientific literacy for two main reasons: (a) statements such as "new ideas are limited by the context in which they are conceived...[would be,] [w]ithout historical examples...no more than slogans" and (b) "some episodes in the history of the scientific endeavor are of surpassing significance to our cultural heritage" (AAAS, 1989, p. 111). Among the examples given are Galileo, Newton,

Darwin, Lyell, and Pasteur. *Science for All Americans* emphasizes "ten accounts of significant discoveries and changes that exemplify the evolution and impact of scientific knowledge" (AAAS, 1989, p. 111). Nine of the ten deal with an important scientific development *and the story is centered on an individual scientist*; the tenth speaks of the industrial revolution and *only one name is cited, very casually*. Could this mean that, in the technological process, individuals are less important than they are in the scientific process? Or that the scientific community is more elitist than the technological community? Or does the difference originate in the ideological biases of the historians who have written that section – or in the dominant ideologies of our culture with respect to science and technology? Could it be possible that science is seen as more ennobling for the individual than technology? If so, why?

The storytellers of science emphasize on the one hand the contribution of individuals and, on the other, the resonance of certain scientific developments with the surrounding culture. The economic, political, and military factors of scientific development are essentially neglected. Another history of science, which would look very different, could be imagined. But, where the history of technology is concerned, the literary genre changes and economics is emphasized.

The kind of history that has been chosen is centered on *results* and neglects the sociohistorical analysis of the development of the *great paradigms* that are at the basis of the scientific processes. One could imagine a history of science that would have tackled questions like: How did people – more precisely, the bourgeoisie – begin to look at things as inanimate and at the world as a *disenchanted world*? Under what cultural, economic, and sociological conditions was the cultural notion of *objective observation* invented? What concept of health developed with the emergence of scientific medicine? What has been the historical development of the concept of scientific law? And so on. This suggests yet another question: "Do the historical perspectives of *Science for All Americans* provide the critical distance one would expect from history?"

Science 5–16 deals with science, but not with the history of science. Even so, some historical perspectives appear in the "social aspects" of science. This report deals with history in a way that is quite different from that of *Science for All Americans*. The interests *Science 5–16* has in history are differences between diverse theories and scientific controversies. The contrast between *Science for All Americans* and *Science 5–16* is consistent with their epistemologies. *Science for All Americans'* epistemology suggests that science observes and describes the world *as it is*. *Science 5–16* has an epistemology that sees science as the human and historical invention of ideas, theories, and models that explain or deal with phenomena. Thus, *Science for All Americans'* history of science will be the tale of the results and the triumph of science, while *Science 5–16* will ask students to "study the ideas and theories used in other times to explain natural phenomena" or "study examples of scientific controversies and the ways in which scientific ideas have changed" (DESWO, 1988, p. 70).

The Concepts of Scientific Literacy

Science Objectives is more interested in assessing scientific knowledge than scientific literacy. Consequently, societal issues are paid little attention. Similarly, *Science 5–16* is interested mainly in a science curriculum. However, at least three characteristics of its proposals are very relevant to scientific literacy: (a) its modern approach to the nature of science with its emphasis on concepts and theories (whereas *Science for All Americans* and *Science Objectives* remain mostly empiricist), (b) the relationship proposed between science and society, and (c) its view of the history of science, one that centers on the workability of former theories and on controversies rather than on people and events.

Science for All Americans develops a complete view of scientific literacy. Its recommendations for what a person should know to be scientifically literate will be the basis for many fruitful debates concerning education and curricula. *Science Education for the Elementary Years* is centered on students and on their growing ability to deal with their environment. Having roots in daily life helps the report develop a sound epistemology, an operational concept of science and technology, and a solid (but not developed) vision of the relationship between science, technology, and society.

Scientific Literacy: An Individualistic Concept?

To conclude, the concept of scientific literacy itself deserves some comment. I propose four orientations related to its ideological assumptions for debate: one, the relationship between the individual's autonomy and corporate interests; two, the role of the experts; three, the limits of the usual concept of observation and even of information gathering; and finally, problems arising from an individualistic approach to scientific literacy. For each of these issues, I submit that, most of the time, the usual conception of scientific literacy avoids confronting the collective and the conflicting dimensions of human life. I would like to show how a collective dimension can be added to scientific literacy *without reducing reality, as Marxists generally do, to conflicts, class struggles, or collectivism.*

When we speak of scientific literacy, whose interests do we serve? Those of the individual or corporate ones? Each report emphasizes both kinds of interests. It seems to be assumed that an invisible hand will always strike a fair balance between the good of each person and the "good of the work force of the nation." This assumption could be challenged. Could it be possible that promoting a scientific culture would not increase the autonomy of individuals, but ultimately would integrate them into a more programmed society in the name of scientific rationality and would subject them to the economic and political interests of the corporate world or—what could be even more disquieting—to the interests of a privileged minority? Even if I do believe that scientific literacy can be beneficial to all, and especially to the poor, I believe that more studies of its social effects are in order.

Related to the previous issue is the question of the role of experts. Individuals in our society very often come into contact with science and technology through experts, like medical doctors, mechanics, engineers, lawyers, architects, physicists, among others. Why is it that so few sections of these reports are devoted to a critical analysis of the relationship between specialists and nonspecialists, especially given that, in "real life," scientific literacy refers to an encounter between experts and laypersons? When people are not experts in a field, they often have to struggle to remain free in the face of scientists. Thus, there is a need for a concept of scientific literacy that would emphasize that being scientifically literate is not only knowing things or being cultivated, but also, implicitly, having a relationship to a specific social power. To put it bluntly, scientific literacy is not simply a disinterested business; power, money, autonomy, and status are all at stake.

In a similar vein, one can wonder if the usual notion of observation, mentioned frequently in these reports and another facet of which was addressed earlier, is conceptualized adequately. It represents, indeed, the gathering of information as a process devoid of conflict. In daily life, that is not often the case. When, for example, I try to learn about the advantages and disadvantages of a commercial product, I must overcome the aggression of the advertising industry. Similarly, a factory worker must struggle to gather information on management policies or on safety issues. These examples show that the usual epistemologies that represent a disinterested observer gathering information truly are limited. Disinterested observation only happens in exceptional situations. In "real life," gathering information usually involves human confrontation—even more, a political process. Scientific literacy, if it wishes to be concerned with "real life," should take into consideration the different circumstances under which observations are made.

Finally, each of the reports presents an individualistic view of scientific literacy. We all know that science is not an individualistic, but a collective, process. Why, then, does the language of scientific literacy not speak of groups? For example, one could speak of the scientific literacy of women, of workers, of specific minorities. In doing so, it would not be enough to say that the goal is to reach women individually, or workers individually, or black people individually. To speak of the collective dimension of scientific literacy, one should take into account that learning is mediated through social institutions, groups, and solidarities. It is through laboratories and departments that scientists research and learn. In a similar manner, it could be that it is mainly through women's organizations, unions, and interest groups that women, workers, and minorities can become scientifically literate? What could be questioned here are the basic assumptions, the basic "habits of the heart," of postindustrialized individualistic society (Bellah, Madsen, Sullivan, Swidler & Tipton, 1986).

Is it possible that the literature on scientific literacy still has to take into account that scientific literacy is not only concerned with knowledge-sharing, but also, and perhaps mainly, with power-sharing? To introduce that dimension into the debate would mean confronting institutional, political, and economic issues that are barely touched on in the four reports.

Is it possible that, in reality, scientific literacy is a societal issue?

Notes

1. An empiricist vision of science could be defined in a popular way as a representation according to which scientific practice starts with an "objective" and "neutral" observation of nature, from which "laws" can be discovered and verified through experiments. An empiricist perspective minimizes the human and historical contexts of scientific development (Fourez, 1988b; Winograd & Flores, 1987) as well as the collective projects that structure scientific processes.
2. All quotations from NCISE, the draft of *Science and technology education for the elementary years: Frameworks for curriculum and instruction*, were used with the authors' permission.

References

American Association for the Advancement of Science. (1989). *Science for all Americans: A Project 2061 report on literacy goals in science, mathematics and technology*. Washington, DC: Author.

Austin, J. L. (1962). *How to do things with words*. Cambridge, MA: Harvard University Press.

Bellah, R. N., Madsen, R., Sullivan, W. M., Swidler, A., & Tipton, S. M. (1986). *Habits of the hearts – Individualism and commitment in American life*. New York: Harper & Row.

Bloor, D. (1976). *Knowledge and social imagery*. London: Routledge & Kegan Paul.

Department of Education and Science and the Welsh Office. (1988). *Science for ages 5 to 16*. London: Author.

Fourez, G. (1982). *Liberation ethics*. Philadelphia: Temple University Press.

Fourez, G. (1985). *Pour une ethique de l'enseignement des sciences* [Toward an ethic of science education/instruction]. Lyons, France: Chronique Sociale.

Fourez, G. (1988a). Ideologies and science teaching. *Bulletin of Science, Technology & Society, 8(3)*, 269–277.

Fourez, G. (1988b). *La construction des sciences* [The construction of the sciences]. Paris: Editions Universitaires.

Habermas, J. (1979). What is universal pragmatics? *Communication and the evolution of society* (pp. 1–68). Boston: Beacon Press.

National Assessment of Educational Progress. (1989). *Science objectives: 1990 assessment*. Princeton, NJ: Educational Testing Service.

National Center for Improving Science Education. (1989a). *Science and technology education for the elementary years: Frameworks for curriculum and instruction* (Draft). Washington, DC: Author.

National Center for Improving Science Education. (1989b). *Science and technology education for the elementary years: Frameworks for curriculum and instruction*. Washington, DC: Author.

Prigogine, I., & Stengers, I. (1988). *Entre le temps et l'eternite* [Between time and eternity]. Paris: Fayard.

Quine, W. V. (1981). *Theories and things*. Cambridge, MA: Harvard University Press.

Ricoeur, P. (1974). Science et ideologie [Science and ideology]. *Revue Philosophique de Louvain, 72*(14), 328–356.

Searle, J. R. (1979). *Expression and meaning: Studies in the theory of speech acts*. Cambridge, England: Cambridge University Press.

Winograd, T., & Flores, F. (1987). *Understanding computers & cognition*. Reading, MA: Addison-Wesley.

5

Views of Scientific Literacy in Elementary School Science Programs: Past, Present, and Future

Morris H. Shamos

A truism in science education is that most elementary school children are captivated by hands-on science activities. Science is basically culture-free, at least at the elementary school level; hence, few children are disadvantaged when it comes to *doing science* in the elementary school grades. The most likely reasons are: (a) at this level, hands-on science depends far less upon reading, writing, and language skills than conventional textbook-based programs, thereby allowing children to feel more comfortable with science, and (b) the child's cultural and family backgrounds have not begun to socialize the child away from the notion that science can be *fun.* Since the 1950s there has been growing evidence that much more can be accomplished at this level than was believed possible in the past: The motivation and ability of children in the primary grades to deal with scientific concepts appear to have been grossly underestimated; many scientists and educators have become convinced that the elementary school years are the time when the greatest impact can be made in science education. This view is strengthened by the belief that many young children have developed their thinking patterns by the age of twelve. Consequently, in these formative years, when children's natural curiosity of about the world around them is at its peak and their minds are so receptive to new ideas, it may be possible to develop a foundation in science that will remain a permanent part of the individual's life – one that will serve students both during their school years and as responsible adult members of society. If little else, it should be possible to avoid reinforcing the misconceptions in science that seem to plague so many students (and adults). The elementary school offers a challenging opportunity for still another reason: with fewer external demands to meet (college entrance requirements, placement examinations, etc.), the elementary school is more receptive to substantial curricular experimentation than the secondary school.

These were among the basic premises on which reform of elementary school curricula in the 1960s and 1970s was based. Such nationally funded programs as Science – A Process Approach (SAPA), Elementary Science Study (ESS), Science Curriculum Improvement Study (SCIS), and Conceptually Oriented Program in Elementary Science (COPES), hailed by science educators then as the modern wave of science education, quickly gained acceptance in a large number, yet a small overall percentage, of elementary schools across the country. Textbooks, teachers' guides, equipment kits, in-service programs, all designed to implement the educational process, became the norm as the science education community sought a more meaningful introduction to science in the elementary schools. A common element in all of these programs was the involvement of children (and teachers) in hands-on activities. In this respect, the new programs were a great departure from most of the traditional elementary school science reading programs then in vogue. Another departure common to these programs was an emphasis on the processes and structure of science. In a sense, the programs sought to represent the scientific enterprise as perceived by the scientist. Indeed, a feature of most curricular reform projects of that period was the extensive involvement of academic scientists in their design; in fact, some science educators claim that the programs were influenced unduly by university scientists (e.g., Office of Technology Assessment, 1988).

It was also in the 1950s and 1960s that the notion of "scientific literacy" as a goal of science education surfaced. Previously, the major goals of science education had a more practical bent, both in terms of curricula and desired outcomes. The post-World-War-II period saw a rapid growth in American peacetime industry, with a corresponding need for many more scientists, engineers, and allied professionals, and, obviously, more science teachers. Therefore, the emphasis was on getting more students into science-related careers. As for the vast majority of students not interested in scientific careers, some exposure to science had long been thought necessary both in high school and college, based on the theory that a discipline so prominent in human affairs deserves to be part of the general education of all students. But true scientific literacy, at least as I (and others) now view it, namely, understanding the principal features of the scientific enterprise, was not the real objective. Instead, the goal was equated somehow with "science for effective citizenship," that is, to develop an informed public capable of playing an intelligent role in science- or technology-based societal issues. Whatever the intended meaning of scientific literacy, which at the time was not clearly defined in an operational sense, it was believed that the new elementary school science programs might lead students toward this elusive goal more effectively than the traditional science (reading) programs then in use (Shamos, 1988a). This failed to occur, as we shall see. Nevertheless, curricular reform continues in the hope that, somehow, a magic formula may be found that will transform the United States painlessly into a nation of scientific literates. The purpose of the present study is to review the content of the current wave of elementary school curricular reform projects (specifically the seven "Triad" or "Troika" programs funded by the National Science Foundation[1]) to determine how they differ from the previous reforms and whether they are more likely to lead to widespread scientific literacy.

Previous Curricular Reforms

Practical science was also prominent in the early part of the 20th century, tempered though it was by John Dewey's admonition that science should be taught in our schools not only because of its practical value but also because it would develop "scientific habits of the mind" in students, that is, such qualities as logical thinking, quantitative analysis, deductive reasoning, proper questioning, and reliance on sound evidence, which would serve the students in all walks of life (Dewey, 1909). Dewey was probably the most famous philosopher of education at the time so it was only natural that science educators would try for the better part of the first half of the century to instill these qualities into generations of students, sometimes under the guise of the scientific method, sometimes by Dewey's rubric, sometimes simply as rational thought – with a virtual total lack of success (Champagne & Klopfer, 1977). Yet, one could well argue that, if faced with a choice, the nation would be better served if the general student actually learned little about science itself, provided he or she managed to acquire the habit of rational thought.

In part, the curricular reforms of the 1960s and 1970s were reactions to the practical science of the previous decades. Beginning in the high schools and eventually moving down to the elementary school grades, developers sought to bring more of the theoretical and conceptual structure of science into the classroom, reflecting the scientific community's view that the facts of science contribute far less toward understanding the nature of the scientific enterprise than the processes of accounting for these facts. Thus, where prior science programs emphasized content (facts, laws, problemsolving, etc.), the new programs stressed the ideas and processes of science (conceptual schemes, theories, analytical reasoning, logic, creativity, etc.). In fact, one of the major criticisms leveled at the new science programs during the 1960s (e.g., Ausubel, 1963; Fischler, 1965; and Atkin, 1966) was that they put too much stress on science process at the expense of content. But, after all, was not this *real science* as seen by the scientist? How else could one portray the essence of this enterprise meaningfully, particularly at a time when the aftermath of World War II, followed by the shock of Sputnik, cast a spotlight on science education? It was the belief that the new programs represented the true nature of science more accurately, coupled with an increasing awareness of the impact of science on society, that fueled the hope that scientific literacy might be an attainable goal. This hope has not been realized, but for reasons that cannot be linked directly to faults of the curricula (Shamos, 1988b).

The millions of students who were exposed to the new hands-on programs of the 1960s have entered the mainstream of society now, prompting the obvious question: Are they more knowledgeable about scientific matters than those who were students during the 1940s and 1950s when science in the elementary school grades mainly involved reading textbooks? In other words, did the carefully designed, "alphabet soup" programs, with their presumed solid grounding in learning theory, science processes, and the foundations of science, have a discernible effect on the students (now adults) who were exposed to them? However more gratifying these new programs may have been to their developers, to the students who were exposed to them at the time, and in many cases to their teachers, there is no evidence, credible or otherwise, that our young adults today are more knowledgeable or sophisticated in science than were previous generations.

Granted that only a small fraction of the elementary school children in the 1960s were exposed to these programs and that most of those who were probably did not follow up with similar science programs in secondary school, the conclusion is nonetheless the same: We see no outward evidence that today's young educated adults are, on average, more literate in science than were previous generations. After all, a major purpose of science education *must* be to produce literate adults, not merely good school performers. If retention or recall of simple scientific concepts eludes the average educated adult, either they were never understood as taught or were lost through disuse, implying that they were not considered important enough to remember.

Unfortunately, no comprehensive longitudinal or retrospective studies are available to quantify this conclusion; hence one must resort to overall impressions from the few available surveys of adult literacy in science, which appears to be no better today than it was a generation or more ago. But there is more telling evidence than this: Had the early reforms been even moderately successful in producing a scientifically literate public, we might not have witnessed the clamor of the 1980s for still another round of reforms in science education. For example, among the many reports of the early 1980s that of the National Science Board Commission (1983) on precollege education in science, mathematics, and technology, urged the development of "new science curricula that incorporate appropriate scientific and technical knowledge and are oriented toward practical issues."

Evaluation of the Earlier Reforms

Perhaps retention into adulthood is not a fair measure of the value of elementary school science curricula. Although, in the final analysis, the mark of a literate individual should be how well he or she fits into and contributes to the good of contemporary society, not merely how well the individual performed in school. At any rate, it may be instructive to look briefly at the results of the standard methods of evaluating innovative curricula as they were applied to the new elementary school science curricula of the 1960s and 1970s; that is, by comparing the performance of experimental groups with that of controls on a contemporaneous basis, namely, shortly after exposure to one of the new curricula. The simple conclusion one can draw is that the new curricula had no profound effect on student performance. Meta-analyses were performed by Bredderman (1983) and by Shymansky, Kyle, and Alport (1983) to synthesize the primary research data obtained from a large number of experimental evaluation studies reported in the literature. Their conclusions are similar: Bredderman summarized his study with the observation that "the overall effects of the activity-based programs on all outcome measures combined were clearly positive, although not dramatically so," (p. 504) while Shymansky et al. concluded that "there is a substantial body of research literature which collectively points to the new science curricula as a successful attempt to improve science education" (p. 402). Such qualified conclusions, taken together with the many assumptions and limitations inherent in both the original studies and the ultimate syntheses, suggest that the overall effects of the new elementary school science curricula must be considered marginal. Whatever positive results were observed, while statistically significant, they are not very impressive. Also, there is always a danger in looking to marginal statistical significance alone, in the absence of other supporting evidence, as the measure of the worth of an educational experiment. Obviously, if a new curriculum exhibited a spectacular effect on overall student performance, it would warrant abandoning quickly the old for the new, but at what level of improvement is it really worth changing? Clearly, elegant as these programs were, the results of the earlier curricular reforms did not impress the science education community sufficiently, to the point of widespread adoption. In fact, there may well be little point in looking to curricular change as the primary means of improving student learning in science. Bredderman found that program differences, on average, accounted for only 5% of the observed variance for all student outcomes combined, whereas if only process outcomes were considered, 10% of the variance could be accounted for.

The finding that process outcomes were more prominent than others should not be surprising because the activity-based programs analyzed by Bredderman (ESS, SAPA, and SCIS) were known to be oriented much more toward scientific processes than content learning. Similarly, in his evaluation of the newly developed high school program, Harvard Project Physics, Welch (1979) concluded that "curriculum does not seem to have much impact on student learning no matter what curriculum variations are used" (p. 301), noting that a greater curriculum impact on students of more than 5% of the total variance was rarely found. When coupled with the fact that science plays such a minor role in elementary school curricula, averaging only about 23 minutes per day (Weiss, 1987) according to teachers' estimates in 1977 (which would have been lower in the 1960s), one probably should not be surprised that the curricular reforms failed to produce dramatic effects. New curricular reform efforts should take due note of these findings for, as Welch points out, other factors such as student ability, time on task, and teacher ability must have played major roles in how well students performed on the assessment instruments used in the studies. To these, one might also add such social factors as family influences, lack of subsequent reinforcement, and (particularly at the secondary school level) peer pressures and simply "growing up."

Another finding from Bredderman's study warrants particular attention in light of our earlier remarks regarding adult literacy. Bredderman reported on three studies designed to examine effects over time of the activity-based elementary school science programs. In the years immediately following elementary school (grades 7, 8, 9, and 10), virtually no lasting effects were observed; in other words, whatever advantages may have been gained during the years of exposure to activity-based science programs appear not to have been sustained in the years immediately following, during which the students were exposed to conventional content-based science programs. Only one outcome, logical thinking, exhibited a small positive influence over time, weighted largely by a study of the effectiveness of SAPA (Raven & Calvey, 1977). It is worth noting in this respect that, in their study of how effectively SCIS developed certain aspects of scientific literacy in children, Boyer and Linn (1978) also found some evidence of long-term effects in two areas that they equate with logical thought processes. But these consequences were also very small. The absence of more positive, long-term improvements is very disappointing, of course, bearing as it does on the seemingly dismal prospect of ever achieving a scientifically literate adult population. It may mean that too much attention was paid to assessing short-term, readily testable results (understandably so) and not enough to assessing long-term retention.

The Meaning of Scientific Literacy

In the past, descriptions of scientific literacy have tended to be rather general and perhaps too broad. In the early days of the current literacy movement, for example, Pella et al. defined the scientifically literate person as one who not only understands the basic concepts and nature of science but also understands the ethics of the scientist, the differences between science and technology, and the interrelationships between science and the humanities and society (Pella, O'Hearn, & Gale, 1966). Clearly, this is a demanding definition that could not possibly be met by the average student (or adult), yet most definitions are of this general nature. Recently, to provide a useful yardstick for his studies on adult literacy in science, Miller (1989) suggested a three-dimensional test of scientific literacy: (1) an understanding of the processes or methods of science for testing our models of reality, (2) a basic vocabulary of scientific and technical terms and concepts, and (3) an understanding of the impact of science and technology on society. There is a good deal of merit in Miller's attempt to codify scientific literacy in this way, but even this begs the question somewhat: For example, are all components equally important or is one component more important than another? What level of "understanding" in (1) and (3) should be taken as the mark of a literate individual? Is some understanding enough or must the individual be able to demonstrate a comprehensive knowledge of both of these aspects of the scientific enterprise to be considered literate in science? If we insist upon the latter we are faced again with a definition that will be impossible to meet, at least in the foreseeable future. Difficult though the task may be, if we are to deal intelligently with the issue of scientific literacy, we must seek definitions that are reasonably specific, bounded, and readily testable.

Few educated individuals are totally illiterate in science; everyone knows some facts of nature and has some conception of what science is about, however naive or misconceived their opinions may be. Thus, it is an oversimplification to assume that one is either totally literate or illiterate in science. Instead, one can distinguish forms or levels of literacy, levels that normally are attained sequentially by students in their formal exposure to science. Following are descriptions (definitions) of three such forms of literacy which build upon one another in degree of sophistication as well as in the chronological development of the science-oriented mind. Because of their vertical structure, I believe they may be more useful as criteria for judging scientific literacy:

1. *Cultural Scientific Literacy.* Clearly, the simplest form of literacy is that proposed by E. D. Hirsch in his recent best-seller. Hirsch (1987) argues that cultural literacy, by which he means a grasp of certain background information that communicators must assume their audiences already have, is the hidden key to effective education. He provides a list of several thousand names, dates, places, events, and so forth, illustrating the type and scope of knowledge that is shared by literate Americans. Included are several hundred, science-related terms that, with appropriate definitions, are claimed to constitute a lexicon for the scientifically literate person; obviously a necessary, but hardly a sufficient, condition for scientific literacy beyond its most primitive meaning. Note the resemblance between this and the basic vocabulary cited by Miller. If all one needed were such a lexicon, scientific literacy would be easy to acquire by rote; such a dictionary, incidentally, has been developed (Hirsch, Kett & Trefil, 1988). Yet, this is the only level of literacy held by most of the educated adults who believe they are reasonably literate in science. They recognize many of the science-based terms (the buzzwords) used by the media, which is generally their only exposure to science, and such recognition probably provides some measure of comfort that they are not totally illiterate in science. But, for the most part, this is where their knowledge of science ends.

2. *Functional Scientific Literacy.* Here, we begin to put some meat on the bare bones of cultural literacy by requiring that the individual not only have

command of a scientific lexicon but also be able to converse, read, and write coherently, using such scientific terms in a nontechnical, but meaningful, context. This means using the terms correctly; for example, knowing what might be called "some of the simple everyday facts of nature," such as having some knowledge of our solar system, that the earth *revolves* about the sun once each year while *rotating* on its own axis once each day, that the moon revolves about the earth once each (lunar) month, and how eclipses occur. Or, to get a little more sophisticated, identifying the ultimate source of our planet's energy, or the "greenhouse effect," or how we get the oxygen we breathe. And, to get still more sophisticated, knowing the difference between electrons and atoms or knowing what DNA is and the role it plays in living things. One could go on and on listing such simple facts about nature, and most objective tests of literacy are based upon just this kind of knowledge or recall. This is not a very demanding requirement, yet estimates of the number of adults in the United States who might qualify at this functional level are distressingly low. Miller (1989) places the fraction of adult Americans who possess a minimal understanding of scientific terms and concepts at about 30 percent. This number seems high to me, but it is clearly a soft number, depending as it does on the sophistication of the test items designed to assess minimal understanding; obviously, one could easily halve this figure by a more rigorous selection. Nonetheless, whatever the number, it does indicate that the scientific knowledge of most adults is very poor, because many of the terms and concepts tested presumably would have been at their fingertips at some point during their schooling. It also shows that whatever exposure they may have to science subsequently through the media fails to reinforce or convey such factual information.

An important difference between cultural and functional literacy, beyond the obvious difference in level of scientific sophistication, is that the former describes a passive state, whereas the latter is more active. That is, not only can the functional literate read and comprehend a newspaper account of a scientific development, for example, as might the cultural literate in some simple cases, but the functional literate also can communicate the substance of that account to a third party, either orally or in writing, in terms that are meaningful to that party.

Determining whether an individual qualifies for one or the other of these forms of literacy (functional literacy presupposes cultural literacy, of course) through objective tests should be relatively easy because both depend largely upon factual recall. But as we have seen, this is not what science is really about. What is lacking in these levels of literacy is an understanding of the processes of science, prescriptions for seeking out and organizing factual information in the unique manner that is characteristic of science, and the fundamental role played by theory in the practice of science. Thus, we arrive at the ultimate or true level of literacy—the one most difficult to attain and to evaluate.

3. *True Scientific Literacy.* At this level of sophistication the individual actually knows something about the scientific enterprise. He or she understands

(or at least is aware of) the major conceptual schemes (the theories) that form the foundations of science, how they were arrived at, why they are widely accepted, how we make order out of a random universe, and the role of experimentation in science. This individual also understands the elements of the so-called "scientific method," of proper questioning, of analytical and deductive reasoning, of logical thought processes, and of reliance upon objective evidence. Obviously, that is demanding, but this is what true scientific literacy ought to mean, which is why we are a nation of scientific illiterates. Some might argue that such a definition is designed to make scientific literacy unattainable, but what it really means is that the term has been used too loosely in the past. It is no great misfortune if true scientific literacy appears to be out of the reach of society at large; most highly specialized knowledge suffers the same fate. Notice that I do not require our scientifically literate person to have at his or her fingertips a wealth of facts, laws, or theories (actually the major conceptual schemes in science can be counted on the fingers of both hands) or be able to solve quantitative problems in science. Nor are advanced mathematical skills essential; all we should expect is that our scientifically literate person understand and appreciate the central role played by mathematics in science. How many of our *science-bound* high school students could qualify for this level of literacy let alone nonscience students? For that matter, how many college graduates could qualify? My estimate of the fraction of Americans who might qualify at this level is 3 to 5% of the adult population, including all professionals in science, engineering, science-related fields, science teachers, science writers, and subscribers to popular science publications. (Miller places this number at about 6% by his benchmarks.)

Having defined these three forms of literacy, the question now is how the definitions may be applied to the evaluation of elementary school science programs and to the potential of such programs for developing scientific literacy in students. The first two (cultural and functional literacy) are easy to evaluate, of course. If we do not set our sights too high, probably most students, by the time they complete high school science, might be considered reasonably literate at the cultural level—and some even at the functional level. But we cannot judge the effectiveness of elementary school science curricula solely by such criteria unless we accept content learning as the sole measure of scientific literacy. Therefore, we must agree on some process or concept goals that are characteristic of true literacy, yet can be measured readily at the grade school level. For this purpose, we must put aside the questions raised earlier about the long-term effectiveness of elementary school science programs or whether, in fact, any curricular changes at this level can have more than a marginal effect on what children learn about science.

All three of the major programs in the earlier round of curricular reform (ESS, SAPA, and SCIS) were hands-on (activity-based), process-oriented curricula. "Hands-on" means that, wherever possible, children were encouraged to base their knowledge of nature upon direct experience; that is, upon observation and experimentation rather than simply reading about it. As for process, the programs sought to put at least as much stress on how one gains and understands information about the natural world as on the information itself. All had assessment phases during which the programs were tested thoroughly. All of them sought to encourage the development of rational thought in the learner, to enlarge the child's understanding of his or her environment, and to develop the child's conceptual structure of science. In short, they were designed to produce changes in children's perceptions of science that would be manifested in behavioral changes. While all sought to convey the scientific community's view of science, only SCIS stated that its primary goal was to develop "scientific literacy" (Karplus & Thier, 1969), meaning both a knowledge of basic concepts and an understanding of the nature of science. More specifically, Karplus regarded the purpose of SCIS as developing in children a combination of basic knowledge of the world and the ability to investigate natural phenomena as well as hoping to instill in them a sense of curiosity about nature (Karplus, 1972). The less widely used program, the Conceptually Oriented Program in Elementary Science (COPES), went beyond SCIS in its attempt to achieve scientific literacy. This was an activity-based program centered on five *great ideas* or conceptual schemes that were believed to portray fairly the basic structure and conceptual development of science (Shamos, 1966). The five schemes are: (a) The Structural Units of the Universe, (b) Interaction and Change, (c) The Conservation of Energy, (d) The Degradation of Energy, and (e) The Statistical View of Nature. Because these conceptual ideas are basic to all of science, it was believed that developing elementary school curricula around so few big ideas would illustrate not only the true nature of science but also might make teachers and students feel more comfortable with science. However, COPES was not as extensively evaluated in the field as were the other programs, particularly with respect to the development of scientific literacy.

A scientific literacy test (SLT) was developed by SCIS based upon specific tasks involving Piagetian-type mental operations that were believed to assess accurately a child's thought processes as well as his or her grasp of the science process involved in a given task (e.g., the concept of variables in an experiment, analyzing experiments, reasoning from data, relative position, histograms, energy transfer, etc.). These are processes that are basic to the practice of science and also encompass, at least in principle, much of what Dewey meant by "habits of the mind." Surely, at the elementary school level, one could not ask for much more by way of an introduction to scientific literacy. Hence, it is discouraging to find that evaluations of the effectiveness of SCIS (using the SLT) showed that, while SCIS students were somewhat more successful on problemsolving tests requiring logical and scientific thinking than were those exposed to traditional science programs, the differences were relatively small (Boyer & Linn, 1978). Nevertheless, in assessing the new "Triad" programs on their potential contribution to scientific literacy we must apply similar criteria, for logical and scientific thinking are clearly among the basic ingredients of such literacy.

The Triad Programs

Almost a quarter of a century after funding the first round of curricular reform projects in elementary school science, the National Science Foundation (NSF) instituted a program to modernize and revitalize elementary school science education in the United States. Presumably, this program was considered necessary because the curricular products of the first round of reforms no longer satisfied the needs of contemporary science education, not because elementary school science had changed, but mainly because making alternative approaches available was felt desirable. Despite their high quality (with respect to their portrayal of science), earlier NSF curricula, as we have seen, failed to produce dramatic results. Neither were they implemented widely in the schools because most of the elementary school teachers were neither prepared nor supported adequately by their school administrators in their efforts to manage these programs successfully. Most teachers were uncomfortable with science, with the logistical problems of managing the activity kits, or both (SRI International, 1987). The summer and in-service teacher-training institutes, while helpful in improving the science background of the highly motivated teacher, failed to capture the interest of the average elementary school teachers, most of whom felt then, as they do today, that their own pre-service education in science was inadequate in preparing them for teaching the kind of science featured in the new programs. And, for the most part, local school administrators, principals, science supervisors, and superintendents failed to develop full support for the programs in their schools. Obviously, teachers and local administrators, but particularly the teachers, are the key to successful education in our schools, be it science or any other discipline. The philosophy, conceptual framework, scope, sequence, and other facets of any new curricula designed for the elementary school grades clearly are directed to the teachers (and administrators), not the students. If, for want of adequate training, the teachers are unable to appreciate fully the essential features and goals of new science curricula, they will not be communicated to the students. In other words, one should not expect scientific literacy to be cultivated in the elementary school by teachers who are illiterate in science.

The NSF's Publisher's Initiatives Program has funded seven elementary school curricular development projects thus far. Each of these so-called "Triad" projects involves a university, a publisher, and a school district. This seems like a sound approach because it ensures the cooperation of a school district in the development and testing of materials and of a publisher for subsequent production and dissemination of the product, the latter having been noticeably absent from the reform movement of the 1960s. Because the publisher must commit substantial financial support to the development phase of the project, his or her interest in the outcome is assured, meanwhile providing greater resources to the developers. On the other hand, while one of the expressed purposes of this form of educational partnership is for curricular innovators to have some influence over what publishers bring into the classroom, there is also concern that the publishers, in seeking a marketable product, may exert a subtle influence on the innovators, in effect endeavoring to avoid radical departures from standard publishing practice. The role of the university (or scientists and science educators) is to provide guidance and control over the scientific content and educational structure of the programs. The projects are in the development stage now, so one can judge only from their stated goals and preliminary materials how they might contribute to the overall objective of the science education community to achieve scientific literacy.

Following are very brief descriptions of each of the programs, including principal features:

- *Science for Life and Living (K-6)* (Biological Sciences Curriculum Study) is an activity-based, integrated science/technology/health program developed in collaboration with the Kendall/Hunt Publishing Company. Discipline concentration is divided equally among the life, health, earth, and physical sciences. The program seeks to integrate the three core areas (science/technology/health) at the end of each year and to incorporate skills in reading, writing, and arithmetic. Developing personal and social goals, problemsolving, decisionmaking, cooperative learning, and technology are emphasized, with in-depth treatment of a few topics rather than a broad-brush overview of many subjects. At each grade level, the curriculum focuses on one major concept and one major skill.

- *Improving Urban Elementary Science (K-6)* (Education Development Center, Inc.) is a hands-on, inquiry-based program, developed with Sunburst Communications, Inc., aimed specifically at urban schools. The project will use teams of teachers to develop about 24, activity-based modules for grades K-6, balancing them among the life, physical, and earth sciences and seeking connections, where possible, with the urban setting. The developers claim that teachers are not intimidated by the program. Basic goals include improving critical thinking, use of language, problemsolving, cooperative group skills, prediction, and verification.

- *Super Science (1-6)* (Scholastic, Inc.) is a mass media program consisting of two classroom science magazines, one for grades 1-3, the other for grades 4-6. Claimed to be a broad-based, scientific literacy program that is "classroom-practical" for teachers, it stresses hands-on, inquiry activities that blend science with mathematics, reading, and social studies to develop scientific and technological skills in children. It also seeks to build attitudes, such as "science is fun," "scientists are real people," "studying science helps us understand how things work," and the like. It attempts to develop process skills and scientific concepts through comic strips, science news notes, directed activities, and stories.

- *Life Lab Science Program (K-6)* (Life Lab Science Program, Inc.) is a garden-based program, designed in cooperation with Addison-Wesley Publishing Company, that attempts to create a context for learning with which children will readily identify; they create their own living laboratory. The project is an expansion of a life science program that has had a decade of trials and piloting. It combines conceptual learning with practical applications to show students through hands-on activities how science relates to everyday life. The scientific themes treated are cycles and changes, energy, structure and function, interdependence, and sustainability. The program also tries to integrate the sciences with other kinds of learning. It is said to provide teachers with friendly, manageable, and affordable hands-on materials.

- *The Science Connection (1-6)* (Houston Museum of Natural Science) is a joint effort with Silver, Burdett, and Ginn Publishing Company. It is a supplementary program for grades 1-6 intended to "expand and enliven" the didactic science text without departing too much from the experience of the reading-based elementary school teacher. The project has three objectives: (a) to produce interesting, student-oriented narrative readers for each grade level, (b) to design related, interactive, hands-on activities packaged in such a manner as to be nonthreatening to the teacher, and (c) to expand science curricula into other content areas and into the community through activities that appeal both to the nonscience-oriented student and the teacher. The program seeks to make connections to mathematics, language arts (vocabulary, etc.), social studies (e.g. the history of electricity), and the fine arts (drawing and music).

- *National Geographic's Kids Network Project (4-6)* (Technical Education Research Center, Inc. [TERC]). These are telecommunications-based curricula, being developed with the National Geographic Society (NGS), which may be used either as supplementary material or as complete year-long science courses in grades 4-6. Known as the "NGS Kids Network," the program will consist of five units in the environmental area, designed to involve students in issues having scientific, social, and geographic significance. Activities include data-gathering at the local school or community level and subsequent transmission of the data to a central computer, which pools the data on a national level and sends back combined results for further analysis. The project is said to focus on a "neglected set of basic skills—skills that foster true 'scientific literacy'." Developers describe these skills as those used regularly by scientists in their everyday work, namely, asking meaningful questions, analyzing problems, collecting, organizing, and analyzing data, using results to answer questions, and so on. Preliminary findings suggest that, generally elementary school teachers are not comfortable with such skills because they lack experience with them.

- *Full Option Science System Project–FOSS (3-6)* (Lawrence Hall of Science) is a collaborative effort with the Encyclopedia Britannica Educational Corporation. It has as its objective the development of "multisensory," laboratory-based activities in the life, physical, and earth sciences. The conceptual design is based upon current cognitive theory, specifically, Lowery's "Biological Basis of Thinking and Learning." The materials are designed to develop scientific process skills and, at the same time, to increase knowledge of vocabulary, language, and social studies. Cooperative learning is employed extensively. Materials include an "Ideas and Inventions" module with the themes of scientific reasoning and technology. The project will have five products: 16 modules of laboratory activities, a procedures guide to help teachers gather equipment and materials for the activities, a set of correlation tables to help integrate FOSS activities into other frameworks (such as textbooks and other programs), laboratory equipment, and worksheets with instructions for the sheets.

Key Features of the Triad Programs

Beyond the novel partnership arrangement for these projects, for which NSF deserves great credit for ensuring ultimate dissemination, they exhibit other common features, reflecting in many cases lessons learned from the first round of curricular reform projects. Chief among them is that they are all hands-on, activity-centered, inquiry-based programs exhibiting the widespread conviction that reading about science is not nearly as meaningful for elementary school children as doing science. All have a more practical orientation than did the first-round projects, aiming to relate science more directly to the students' everyday experiences, including a discernible trend toward technological topics. All employ cooperative learning as an instructional tool. A common tendency is to cover fewer concepts, but in greater depth; also there is an effort to make the programs less demanding for the teacher to use (in terms of the boxes, kits, and bits and pieces that add to the logistical complexity of activity-based science programs). All of the programs stress scientific processes and seek to integrate science with other learning, the latter on the sound premise that, by doing so, teachers and school administrators will find the programs more attractive, thereby increasing the time devoted to science-related activities. And, of course, all claim to apply current insights in cognitive development.

A characteristic of all of the Triad projects is the absence of academic scientists in leadership roles. Instead, unlike the earlier round of reforms, which were guided mainly by scientists in association with science educators, these are directed by science educators with scientists as consultants and advisors. It should be noted that the control of curricular reform projects by scientists, rather than by science educators, has been criticized in the past as improperly influencing the direction taken by such programs in regard to scientific content and structure (Office of Technology Assessment, 1988). Certainly, it is fair to say that many of the previous programs, notably SCIS and COPES, are unusual not only because of their insistence upon scientific accuracy – which obviously must be a goal of all science programs – but also, as pointed out earlier, for their attempts to present science as scientists perceive it. This leads to the final, and perhaps most significant, observation to be made about the Triad programs.

The first round of curricular reform programs, as we have noted, were too complex for the average elementary school teacher, both in terms of the science involved and in the logistics of handling the equipment kits. Without fairly extensive, in-service training, few teachers felt comfortable with the programs; even with such training, many still had doubts about their ability to deal with the more subtle concepts of science. And, today, very few of the teachers who were trained in the first round remain in the schools, thereby compounding the problem. The fact is that the pre-service education of prospective elementary school teachers fails to prepare them adequately to teach science along with everything else they must do in the classroom. A few weeks of in-service training, if at all effective, is hardly the ideal remedy for this well-known problem.

The teacher problem appears to have been well recognized by all of the Triad developers. One finds in the project descriptions such expressions as "teachers are not intimidated," "classroom-practical for teachers," "user-friendly (to teacher) materials," "teachers feel comfortable with the program," and so on. The TERC project, for example, reports that it found some teachers uncomfortable with data analysis skills, which is not at all surprising, and that this has influenced greatly TERC's plans for revisions and design of other units. The question is whether in trying to make these programs "teacher-proof," the science must be watered down to the point where the teachers will fail to see science as it really is. There ought to be an axiom of education that, at some point in a child's development, teachers cannot teach what they do not understand fully themselves. Where this point may be in science education is not clear, but certainly at some grade level in the elementary school the effective teacher *must* understand more about science than can be conveyed through teachers' guides and in-service institutes. This poses manifold questions about science in the elementary schools that go beyond the main purpose of this paper – questions about the pre-service and in-service science education of elementary school teachers; whether, in fact, it is possible to achieve a reasonable level of scientific sophistication in roughly 1.5 million elementary school teachers in the United States, or whether, ultimately, we will need science specialists in our elementary schools. All of this bears directly on the basic question that prompted this review—that of scientific literacy.

The Triad Programs and Scientific Literacy

The question of whether the Triad initiative will promote scientific literacy in elementary school children invites a complex array of answers. If the question is posed relative to the earlier endeavors at curricular reform, that is, will the Triad programs be more successful than their predecessors in developing *true scientific literacy* (as defined above), the answer is an emphatic no. Because the earlier programs were not very successful in this respect, the Triad projects, lacking emphasis on the processes and foundations of science, are even less likely to succeed. The question is not whether these projects will produce a more literate adult population, but whether the subject matter of the Triad undertaking can be viewed as contributing directly to the scientific literacy of the students even on a contemporaneous basis. It is in this respect that the answer must be negative, for the curricula under discussion do not stress the most fundamental quests of science, namely, how to learn that the laws and theories (the *big ideas*) that science has evolved may be used to account for the observed facts of nature. It is wrong to assume, for example, that some understanding of the concept of energy, or even of how energy may be transformed from one form to another, makes one fully literate on the subject of energy, for without an appreciation of the overarching conceptual scheme of the *conservation of energy,* the student (and teacher) are left with a substantial void in their conceptions of science.

This comment is not meant as adverse criticism. In fairness, it must be noted that the Triad programs on the whole do not pretend to have scientific literacy as a primary goal. The two Triad projects that do claim scientific literacy as a target *(Super Science* and *NGS Kids Network)* use the term loosely, that is, in the generic sense rather than the meaning of true scientific literacy as defined here. Indirectly, all science education, particularly inquiry-based learning, may be said to contribute to the ultimate goal of scientific literacy, but it is a mistake to believe that young students could synthesize, on their own, the pieces of any of the Triad programs into a comprehensive view of the scientific enterprise. The programs should result in some degree of cultural, and perhaps even functional literacy, as these stages have been defined, but as for true scientific literacy, the best results that can be hoped for, I believe, are a liking for science (or at least not a loss of interest), perhaps a better understanding of how science goes about its business, and, intuitively some of Dewey's "scientific habits of the mind." Should these prove to be outcomes of the Triad projects, they would be no small accomplishments and would surpass by far any previous achievements in elementary school science. To the extent that such progress may open the door for some students to attain true scientific literacy later on, or encourage more students to pursue science or engineering careers, the Triad programs will have done all that can reasonably be expected from elementary school science.

Conclusion

True scientific literacy, as an immediate goal of elementary school science, is an unreasonable expectation—but not so much for want of good curricula. Almost any activity-centered, inquiry-based program, in the hands of a qualified science teacher—meaning one who is reasonably literate in science and who engages students for more than the bare minimum of time normally allotted to elementary school science each week in the United States—should produce impressive results. As we know, however, very few elementary school teachers, through no direct fault of their own, are entirely qualified to teach science. Hence, either we must continue to design elementary school science programs to be as teacher-proof as possible (which appears to be a guiding principle in the Triad programs), in which case we should lower our sights from true to (at best) functional scientific literacy, or society must determine that science education is important enough to justify the expense of having trained science specialists in all of the nation's elementary schools. Until that stage is reached, elementary school science would do well to focus on developing "scientific habits of the mind" and a fuller appreciation of science.

Notes

1. "Triad" and "Troika" are popular names given to the National Science Foundation-sponsored Publishers Initiative Program because the programs involve a partnership of a university (or scientists and science educators), a publisher, and a school district. Programs funded under this initiative thus far are: (1) *Science for Life and Living*, Biological Sciences Curriculum Study (Principal Investigator: Rodger W. Bybee); (2) *Improving Urban Elementary Science: A Collaborative Approach*, Education Development Centers, Inc. (Co-Principal Investigators: Karen Worth and Judith Sandler); (3) *Super Science: A Mass Media Program*, Scholastic, Inc. (Principal Investigator: Victoria Chapman); (4) *The Life Lab Science Program: Development of a Comprehensive Experimental Elementary Science Curriculum*, Life Lab Science Program, Inc. (Co-Principal Investigators: Gary Appel and Roberta Jaffe); (5) *The Science Connection*, Houston Museum of Natural Science (Co-Principal Investigators: Carolyn Sumners and Terry Contant); (6) *National Geographic's Kids Network Project*, Technical Education Research Centers, Inc. (Principal Investigator: Robert F. Tinker); and (7) *Full Option Science System—FOSS*, Lawrence Hall of Science (Principal Investigator: Lawrence Lowery).

References

Atkin, J.M. (1966). Comments on two approaches. *Science, 151,* 1033-1055.

Ausubel, D.P. (1963). Some psychological considerations in the objectives and design of an elementary school science program. *Science Education, 47,* 278-284.

Boyer, J.B. & Linn, M.C. (1978). Effectiveness of the Science Curriculum Improvement Study in teaching scientific literacy. *Journal of Research in Science Teaching, 15*(3), 209–219.

Bredderman, T. (1983). Effects of activity-based elementary science on student outcomes; a quantitative synthesis. *Review of Educational Research, 53*(4), 499–518.

Champagne, A.B. & Klopfer, L.E. (1977). A sixty-year perspective on three issues in science education. *Science Education, 61,* 437–46.

Dewey, J. (1909). Symposium on the purpose and organization of physics teaching in secondary schools; part XIII. *School Science and Mathematics*, 9.

Fischler, A.S. (1965). Science, process, the learner: A synthesis. *Science Education, 49(3)*, 402–409.

Hirsch, E.D., Jr. (1987). *Cultural literacy: What every American needs to know.* Boston: Houghton Mifflin Company.

Hirsch, E.D., Jr., Kett, J.F. & Trefil, J. (1988). *The dictionary of cultural literacy.* Boston: Houghton Mifflin Company.

Karplus, R. & Thier, H.D. (1969). *A new look at elementary school science.* Chicago: Rand McNally and Company.

Karplus, R. (1972). Physics for beginners. *Physics Today, 25*(6), 36–47.

Miller, J.D. (1989, January). *Scientific literacy.* Paper presented at the 1989 Annual Meeting of the American Association for the Advancement of Science, San Francisco, CA.

National Science Board Commission. (1983). *Educating Americans for the 21st century* (CPCE-NSF-03). Washington, D.C.: National Science Foundation.

Office of Technology Assessment. (1988). *Elementary and secondary education for science and engineering* (OTA-TM-SET-41). Washington, D.C.: U.S. Government Printing Office.

Pella, M.O., O'Hearn, G.T. & Gale, C.W. (1966). Referents to scientific literacy. *Journal of Research in Science Teaching, 4(2)*, 199–208.

Raven, R.J. & Calvey, Sister H. (1977). Achievement on a test of Piaget's operative comprehension as a function of a process oriented elementary school science program. *Science Education, 61(1)*, 159–166.

Shamos, M.H. (1966). The role of major conceptual schemes in science education. *The Science Teacher, 33*(1), 27–30.

Shamos, M.H. (1988a). A false alarm in science education. *Issues in Science and Technology, 4*(3), 65–69.

Shamos, M.H. (1988b). The lesson every child need not learn. *The Sciences, 28(4), 14–20.*

Shymansky, J.A., Kyle, W.C. & Alport, J.M. (1983). The effects of new science curricula on student performance. *Journal of Research in Science Teaching, 20(3)*, 387–404.

SRI International. (1987). *Opportunities for strategic investment in K–12 science education: Options for the National Science Foundation* (SRI Project No. 1809). Menlo Park, CA: Author.

Weiss, I. R. (1987). *Report of the 1985–86 national survey of science and mathematics education* (RTJ/2938/00–FR). Chapel Hill, NC: Research Triangle Institute.

Welch, W.W. (1979). Twenty years of science curriculum development: A look back. In D.C. Berliner (Ed.), *Review of Research in Education*, Washington, D.C.: American Educational Research Association.

6

Elementary School Science Curricula That Have Potential to Promote Scientific Literacy (And How to Recognize One When You See One)

Angelo Collins

Selecting elementary school science curricula that promote scientific literacy—scientific knowledge, skills, and dispositions—is a difficult task. The educator must assess the congruence between the outcomes claimed by the curricula developers and the outcomes that will result from actual implementation of the curricula in light of the goals the teacher or the school district has for the students. This chapter is designed to provide teachers, school administrators, district and state level curricula coordinators, and others with insights and guidance for assessing elementary school science curricula.

At a minimum, curricula that promote scientific literacy should provide elementary school children with opportunities to:

- observe and describe natural events
- pose questions about natural events
- explain natural events using scientific terms accurately
- recognize that the concepts that support those terms are human inventions and not immutable truths

- develop the skills required to describe, explain, and predict natural phenomena
- design experiments that test predictions about natural events
- develop the skills to work with other students to produce scientific descriptions, explanations, and predictions
- appreciate that scientific knowledge is constructed by people for people to describe, explain, predict, and control natural phenomena; and as such, it is complex and subject to change
- be comfortable enough with the ideas and procedures of science to follow a debate about a scientific topic in the media
- be aware of the influence of scientific knowledge on daily life and of daily life on science
- develop a lifelong enthusiasm for and excitement about knowledge of nature.

Curricula that provide these opportunities have the potential to prepare students to function in a scientific and technological society. These goals are the standards for the method of assessing curricula developed in this chapter.

Nothing in this chapter is new or revolutionary. Rather, it is an overview of some issues in science education and curricula development and selection, with some suggestions on how these issues relate to the problem of achieving the best match of elementary school curricula to the goals and students of a particular school or district. The chapter is based on the assumption that one goal of teaching elementary school science is and should be developing an appropriate level of scientific literacy for all children. The chapter begins with operational definitions of science, scientific literacy, and curricula and then continues with a description of a process for examining a curriculum for elements of scientific literacy. The process or model is then applied to three of the seven elementary school science curricula that were developed recently with support from the National Science Foundation.[1] (These curricula are joint efforts of groups doing research and development in science education and commercial publishers.) As the reader will discover, despite the best intentions of the research and development groups, the publishers, and the federal government, no one curriculum will match a particular school system's requirements exactly. Such a curriculum, developed at the national level, cannot be suitable for all school districts. The chapter concludes with some caveats.

Definitions

Before a science curriculum can be assessed as a whole, it is important that its components be defined clearly. Operational definitions facilitate the objective assessment of the relationships among the developers' claims, the developers' recommendations for implementation, and the actual achievement of goals. The definitions serve as guideposts. They signal the presence or absence of key concepts in a curriculum, as well as the context and correctness of the use of these concepts. Assessing elementary school science curricula requires an operational definition of the nature of scientific understanding—the knowledge, skills, and dispositions that characterize science and the scientific enterprise. The nature of scientific literacy is rooted in these contexts.

The Nature of Science

What is science? There are probably as many different definitions of the nature of science as there are people reading this chapter. One definition is that science is knowledge constructed by humans to describe, understand, explain, predict, and control natural phenomena. One way to look at the complex entity encompassed by scientific knowledge is to look at it as if it had three discrete components: structural, procedural, and human. Let us explore each component separately.

Structural Component. First, let us examine the structural component of scientific knowledge—the content—the knowledge products of scientific inquiry and the events on which these knowledge products are based. The elements in the structure of scientific knowledge are analogous to the materials needed to construct a building. The elements are events, facts, concepts, relationships, theories, and models. Events are observed: an orange left in a student's locker for weeks becomes moldy; a switch on the wall is manipulated and electric lights glow overhead; a sand castle left on the beach at night is obliterated in the morning. Events are the beginning, end, and reason for science. Facts are empirical statements that report observations of events. When Maria reports to Mr. Yee that she left an orange in her locker during the spring break and now it is yucky, that is a statement of fact—a report of her observations of an event. Mr. Yee, being a true science teacher, may ask Maria to describe the look, the smell, and the feel of the moldy fruit. He is asking Maria to use several of her senses, not just sight, to make her observations and to report as many facts as possible. He is providing Maria with an opportunity to experience science.

Concepts are the mental representations that Maria and Mr. Yee, as well as you and I, have that give meaning to and receive meaning from the facts. The connection between facts and concepts is reciprocal. Events and the facts that are reported about them help to build the mental representations that are concepts, which, in turn, are used in stating facts to report events. As Maria observes the dull olive green circle of fuzz which she calls "yucky," but Mr. Yee calls mold, she has experienced one event to begin to build her concept of mold. Concepts may be empirical—the result of observation—or abstract and theoretical—products of the imagination that may or may not exist in a concrete sense but make explanation and prediction possible.

Concepts are idiosyncratic, the results of each person's experiences of events, but, as people generally experience similar events, there is sufficient similarity in their concepts to make communication possible. Conceptual structures may be poor, the result of only one or two events, or rich, built from many years' experiences. Maria's conceptual structure of the moldy fruit may consist of an orange, green and fuzzy, with a pungent odor. Mr. Yee's conceptual structure may consist of organic matter including oranges, bananas, and human flesh, spores including a mycetomatous life cycle, growth conditions including food, moisture, darkness, and colony descriptions, and an economic aspect including both athlete's foot and penicillin.

As relationships among concepts become helpful in describing more and more events, scientists label such relationships principles or laws. Hypotheses, statements that both explain some events or some aspect of an event and predict others, are essential in the work of scientists as they design experiments to test their predictions. A conceptual structure that is rich with many relationships and principles useful in explaining several related events is called a theory; another structure that is rich with useful relations to predict natural events is termed a model. Events, facts, concepts, relationships, theories, and models are the building blocks from which scientific knowledge is constructed.

Procedural Component. Just as materials are necessary, but not sufficient, to construct a building, facts, concepts, relationships, theories, and models are necessary, but not sufficient, to construct scientific knowledge. As a builder has ways of manipulating materials, a person constructing scientific knowledge requires a repertoire of skills to manipulate the elements of scientific knowledge – the procedural knowledge of science. There have been many attempts to describe, define, and delineate the procedural knowledge of science, commonly called "the scientific method." Textbooks tell us that scientists make observations, form hypotheses, experiment, and draw conclusions to make theories (Alexander, 1986), or alternatively, that they define a problem, collect background information, formulate a hypothesis, test the hypothesis, make and record observations, and draw conclusions (Goodman, Emmel, Graham, Slowiczek & Schecter, 1986). Although these statements may be true in some instances, such linear descriptions do not capture the complexity and serendipity involved in building scientific knowledge.

Delineating "process skills" or "inquiry skills" is another attempt at a different level of detail to define the repertoire of ways a scientist manipulates the structural elements to build scientific knowledge. Lists of categories of procedural knowledge include: observe and describe, compare and contrast, classify, infer, measure, manipulate and control variables, interpret data, define operationally, formulate hypotheses, communicate, and make predictions. Variations on the list might include analyze and synthesize or recognize patterns.

The current emphases in the curriculum on critical thinking, problemsolving, and creativity are attempts to identify and describe the mental manipulations required to build conceptual structures in science or in any discipline. From a critical thinking stance, one might adopt the position that the procedural knowledge of science includes analysis, synthesis, inference, deductions and metacognition. From a problemsolving stance, one might say that the procedural knowledge of science includes the ability to recognize a problem and/or pose a question, identify the essentials of the problem, attempt a solution, and check the solution for accuracy and completeness. From a creativity stance, one might describe procedures to observe familiar events in new ways, which is surely one method for building scientific knowledge. The procedural knowledge of science, then, may be defined as any activity, in which a person engages, that involves the manipulation of the elements of structural knowledge and builds conceptual structures.

If these several attempts to define the procedural knowledge of science are not sufficiently confusing, there is yet another type of procedural knowledge – the use of tools. Building scientific knowledge relies on the use of both material tools, from meter sticks to computerized spectrophotometers, and mental tools, from accurate estimation to employing heuristics and algorithms.

Human Component. There is one more component to scientific knowledge – the human component. Contrary to the stereotype that persists in comics, cartoons, and other media, science is not a solitary enterprise. The interrelatedness and human component of science can be found in the relations of the natural sciences to each other and to other disciplines and in the relations of scientific knowledge to daily life. Despite the way it is often taught in school, natural science is not compartmentalized into biology, chemistry, physics, and earth science. These categories are selected arbitrarily in order to simplify the complexity of science for instructional purposes. To study any natural phenomena, scientists learn several areas themselves and then work with and rely on specialists in other areas. Also, science is not isolated from reading, writing, mathematics, or social studies. Scientists employ reading, writing, and mathematics to communicate the knowledge they have constructed. The history of an idea is often as important as its current status. Nor is science isolated from aesthetics; simplicity of explanation or design is one criterion for a successful theory.

An intimate and mutual connection also exists between science and daily life. What scientists come to know as they attempt to build knowledge structures that explain and predict natural phenomena exerts influences on and is influenced by the daily life of persons in society. For example, the development and acceptance of the *germ theory* of disease influenced health care and lifestyle in the latter half of the 19th century. Also, the concerns of society influence the events and problems that scientists choose to explore. The energy crisis is as much an impetus to the search for cold fusion as is the quest to demonstrate that such energy does exist. Daily, we are consumers of the products of science – whether it is deciding where to place a radiator for the best energy use or practicing with a Frisbee to develop better throwing ability. We regularly ask questions about natural phenomena that influence the work of scientists – how can we limit the erosion on a beach or what shall we do about the greenhouse effect? Science not only proposes solutions to some of society's problems but also it creates other problems. For example, through science, we have learned to control hemophilia – an achievement that has improved the quality of life for many people – but we have increased the probability of women having this genetic disease. The dynamic between science and society is inseparable.

Because scientific knowledge is a human construct, it is tentative. This knowledge is useful only as long as it is able to explain and predict. At any time, an event may occur that cannot be explained by current scientific knowledge. Then scientific knowledge needs to be restructured and reorganized. Or technology may provide scientists with tools to make more precise observations, to record more extensive data, or to analyze it more quickly. This new information results in the restructuring of scientific knowledge.

Any attempt to break science into components destroys its integrity. However, the failure to recognize at least these three components of the scientific enterprise restricts both teachers and students to a concept of science that does not capture its complex, dynamic, and human character. Teaching science without a balanced emphasis on the conceptual, procedural, and human components may be teaching something, but it is not science. A balanced approach to science provides students with the opportunity to develop the skills to construct and evaluate their own knowledge as well as to develop knowledge structures. Science may be as abstract and esoteric as the description of the behavior of a photon or as commonplace and mundane as a child making and flying different styles of paper airplanes. But, in either instance, science is knowledge by and for humans. Perhaps it is not stretching an analogy unduly to say that science is like a three-ring circus. It is usually not possible to attend to all three rings at once, but the experience is diminished if all three rings are do not receive attention at some time. Figure 1 is a display of scientific knowledge as a three-ring circus – the rings are not placed one on top of the other, because then it is difficult to distinguish any one component. They are not completely separate, so that the interrelationships of the components are not lost. This heuristic will be helpful later when we examine elementary school science curricula to see how they present the nature of scientific understanding.

After our extended discussion of the nature of science, the question remains: what is the relationship among the nature of science, scientific literacy for elementary school children, and curricula that support such scientific literacy? Do elementary school students need to be able to distinguish between structural and procedural knowledge to be scientifically literate? Of course not. Do elementary school teachers need to be able to distinguish a theory from a model? Probably not. Do developers of curricula guides need to consider the nature of science in their design? Definitely.

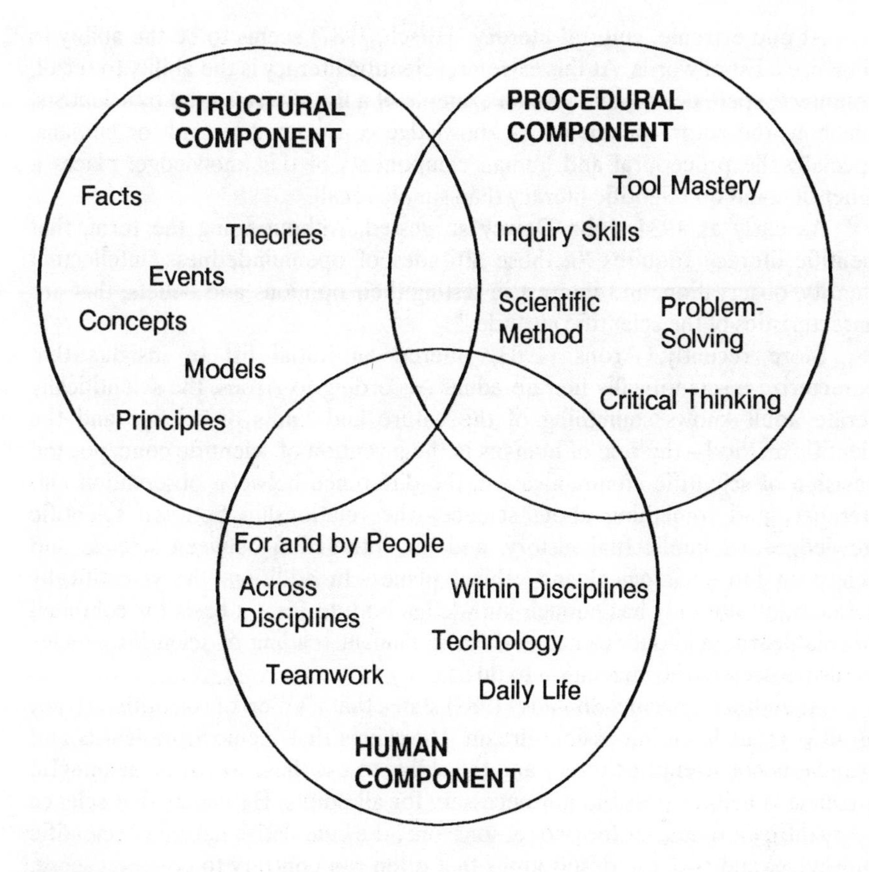

Figure 1
Components of Scientific Understanding

Scientific Literacy

Scientific literacy is a term that is frequently used but seldom defined, on the assumption that everyone knows what it means. Let us look at several attempts to define scientific literacy. Literacy is defined as the ability to communicate – to read, write and speak – at a functional level. Inherent in the definition of literacy is the idea that the ability to read, write, and communicate empowers a person. Would scientific literacy, then, be the ability to read, write, comprehend, form opinions, and discuss intelligently historic and current problems in the natural sciences? This would seem to be an impossible and unrealistic task. Even those who claim science as a profession are seldom able to carry on such discussions outside their own specialties and closely allied areas. What, then, is the power associated with scientific literacy?

At one extreme, cultural literacy (Hirsch, 1987) seems to be the ability to recognize a list of words. At this extreme, scientific literacy is the ability to recall, pronounce, spell, define, and give an example of a list of terms used by scientists. Somehow, the nature of science as knowledge constructed by and for humans, especially the procedural and human components of this knowledge, places a higher demand on scientific literacy than simple recall.

As early as 1934, John Dewey suggested, without using the term, that scientific literacy requires "...those attitudes of openmindedness, intellectual integrity, observation, and interest in testing their opinions and beliefs, that are characteristics of the scientific attitude."

More recently, Arons (1983) offered an initial list of insights that characterize a scientifically literate adult. According to Arons, the scientifically literate adult knows something of the nature and limits of science and the scientific method – the role of humans in the invention of scientific concepts, the precision of scientific communication, the difference between observation and inference, and something about science – the relationship between scientific knowledge and intellectual history, and the interaction between science and society on the sociological and ethical planes. In addition, the scientifically literate adult not only has enough knowledge both to form a basis for continual informal learning about science and for intelligent reading of scientific articles but also possesses the inclination to do so.

At another extreme, Shamos (1988) states that a vision of scientific literacy for all persons is an impossible dream. He claims that memorizing words and formulas is not scientific literacy and the ability to use these words in meaningful discourse is neither possible nor necessary for all adults. He claims that science is very difficult to master for two reasons: one, the cumulative nature of scientific knowledge, and two, the descriptions that often run contrary to common sense. He proposes, but does not delineate, an educational goal of scientific appreciation, analogous to music or art appreciation. Just as all students will not become professional artists, yet all students can appreciate the procedures used by artists and the work they produce. Likewise, not all students will become professional scientists, yet all can appreciate what scientific knowledge is, how it comes to be, and that it is by and for humans.

In early 1989, the announcement advertising the American Association for the Advancement of Science's (AAAS's) Forum 89 asked the recipients to rank 15 components of scientific literacy, with regard to the abilities that should characterize the typical high school graduate. Among the 15 components are the abilities to pose a question that can be addressed by scientific methods, to give a scientific explanation of a natural process, to use appropriate methodology, to read and understand science as it is presented in a newspaper, to interpret graphs, to envision science as worthy of pursuit even without immediate practical gains, to define terms accurately, to design an experiment that is a valid test of a hypothesis, to describe natural phenomena, to use science in personal decisionmaking, and to locate information about science when needed.

Let us assume, as a minimal goal, that a child in elementary school who is in the process of becoming scientifically literate has opportunities to develop the abilities that were listed at the beginning of this paper. Even minimal goals such as those have the potential to enable students to function in a scientific and technological society. However, what is still missing from all of the attempts to define scientific literacy is an enthusiasm and excitement for knowledge about science.

Curricula

What would curricula look like that would promote such a vision of scientific literacy for all elementary school children? Before we attempt to answer such a question, we need some agreement about the definition of curriculum. A science curriculum is the result of decisions about what to teach. For example, from all of the things that might possibly be taught to children age 7 and called science, what is the most appropriate? Several important factors influence the decisions about what will be included in their curriculum. These include: a conception of the nature of science, the problems, needs, and interests of the students who will learn from the curriculum; the concerns of the community in which these students reside, a theory of how students learn science, and the goals of the curriculum. Considering these factors, it is not surprising that curricula are revised regularly.

First, because scientific knowledge is a human construct, it is constantly being discovered, revised, and discarded. Research questions are redefined and consequently the technology to assist research improves. In this century, some elementary school science curricula have emphasized solely the structural component of science, the content curricula; others have emphasized only the procedural component of science, the process curricula; and still others were concerned exclusively with aspects of daily living, the social functions curricula. One hopes that the curricula being developed have a balanced approach to the nature of science.

Second, the problems, interests, and needs of children change. Some of the interests of children are determined by geography—children living in the city may have interests that differ from those of children in rural areas. Other interests are determined by age, social pressures, or local events.

Third, the concerns of the society in which the children live also change. As the most obvious example, five years ago there was no reason for curricula to address issues related to AIDS.

Fourth, the educational community's understanding of how people learn is based on theoretical knowledge, attempts to explain and predict the phenomena associated with learning—remembering, forgetting, reasoning, solving problems, and being creative. As such, learning theories, like scientific theories, are human constructs. They are considered powerful when they explain and predict many different phenomena in various contexts; they are abandoned when they lose their power. For example, one learning theory that has had much influence on science curricula in this century is behaviorism. Briefly and extremely, it states that no one is able to know what happens in the mind, so the best that teachers can do is measure the behaviors that are the result of learning. Hence, curricular designs that evolved from a behaviorist learning theory contained learning objectives written as behaviors that students should be able to demonstrate after instruction.

Constructivism is a theory that is currently proving to be very powerful in explaining and predicting the phenomena of science learning. As Bybee (1988) stated in the 1988 volume of *This Year in School Science*:

> Constructivism is a dynamic and interactive model of how humans learn. According to this theory, students redefine, reorganize, and elaborate their existing concepts through interactions with objects, peers, and events in their environment. Students "interpret" objects and phenomena and subsequently explain their world in terms of their current conceptual understanding. Application of constructivist theory to teaching involves challenging students' current conceptions. (p. 160)

As scientific knowledge is constructed (a philosophical position about the nature of science) and as students learn by constructing knowledge (a psychological position on learning), there is growing evidence of an affinity between constructivism as a learning theory and science learning.

The fifth and final factor influencing decisions about what will be incorporated into a curriculum is the objective of the curriculum. The assumption of this chapter is that the primary goal of contemporary science curricula is scientific literacy, as defined above, for all children.

Considering the five factors that influence decisions about elementary school science curricula, it is not surprising that curricula are written by committees with members such as parents, teachers, specialists in scientific disciplines, educational psychologists, members of the community, and science educators. Nor is it surprising that nationally prepared curricula do not meet the needs of local school districts. The members of the committee that designed the national curricula cannot address both the specific needs of a rural community faced with problems of drought and the needs of an urban community concerned with issues of overpopulation. The study of a food web might be an appropriate conceptual structure for students to learn, but it would be expected that the examples that will be meaningful to students in Kivilina, Alaska will be different from the meaningful examples for students in Key West, Florida. Further, the committee that designed the elementary school science curricula knows nothing of the diversity in any one teacher's classroom in any given year. Therefore, it is more appropriate that curricular documents be called curricular guides.

Another reason that the materials a teacher receives when elementary school science curricula are ordered do not truly constitute a curriculum is that the materials typically include suggestions, directions, or prescriptions for instruction. Instruction is a separate decision; it is the answer to the question which method to use to achieve the goals set by the curriculum. Although intimately tied together, curriculum and instruction are separate issues. Determining that students in the third grade should know about series and parallel circuits is a curricular decision. Students might achieve this goal by listening to a teacher talk, doing a worksheet, reading from a book, observing a demonstration, watching a movie, or manipulating batteries, wires, switches, and light bulbs, or bashing a hole in the classroom wall to observe the wiring. These student activities might be done alone, with a partner, in teams, or in cooperative learning groups. The instructional suggestions in a curriculum guide should be for methods of instruction that are consistent with the learning objectives of the curriculum. In addition to the curriculum, there are many factors that influence the decisions about instruction – the cost of materials, potential safety hazards, class size, and time, for example. Therefore, a purchased curriculum usually contains statements of rationale (what to teach and why) and a series of lessons (how to teach it). Figure 2 is a diagram of the factors involved in making

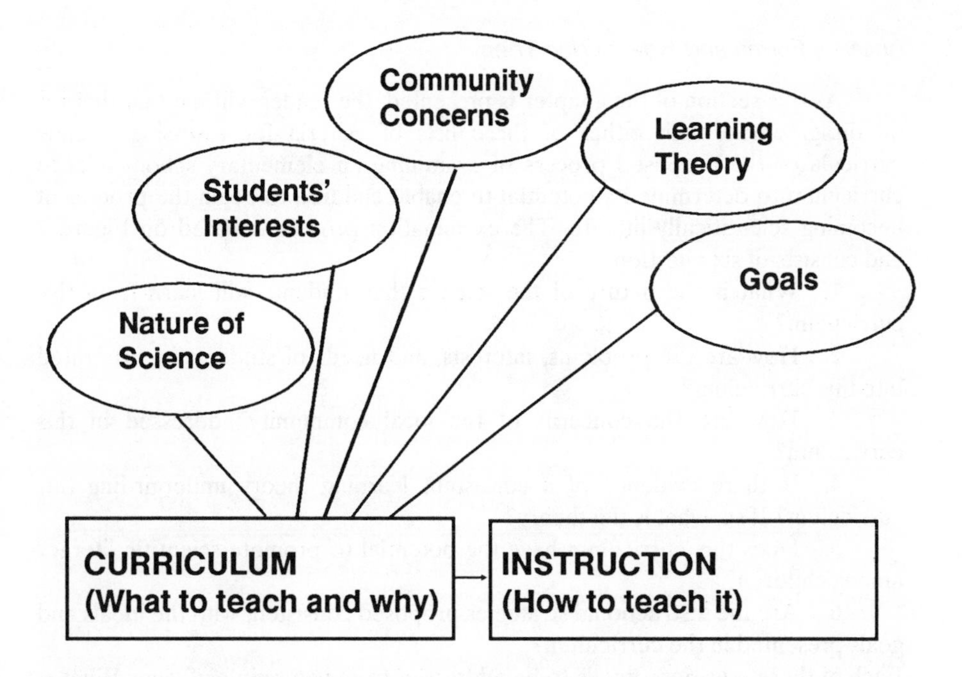

Figure 2
Factors that Influence Decisions About the Curriculum

decisions about curricula. It also will serve as a useful guide when we look at a process to examine a curriculum for its potential to promote scientific literacy.

A Process for Examining an Elementary School Science Curriculum

Setting criteria to evaluate science curricula is not a new task. In 1985, Raizen and Jones proposed looking at textbooks as the prime source of science curricula and analyzing them along the following lines: balance between the learning of recorded knowledge and its application, emphasis given to specific important topics, adherence to the logic of the discipline, opportunity and guidance for students to discover knowledge, and incorporation of a learning theory. For the 1986 AAAS Forum, Raizen proposed four indicators of a quality science curriculum: (a) content coverage, (b) depth of treatment, which includes "basic understanding and appreciation of the structure of the scientific discipline, the process of doing it, and some of the complex problems solved and created by its application" (p. 197), (c) scientific accuracy as found in texts, teacher guides, and/or laboratory materials, and (d) pedagogic quality, which includes instructional strategies, designs, and sequences.

Analysis Forms and How to Use Them

As this section of the chapter is presented, the reader will see that there is no disagreement with either of these lists of criteria for assessing science curricula or the proposed process of examining an elementary school science curriculum to determine its potential to enable children to begin the process of becoming scientifically literate. The examination process is based on Figure 2 and consists of six questions.

1. What is the nature of the science that students will learn from this curriculum?
2. How are the problems, interests, and needs of students incorporated into this curriculum?
3. How are the concerns of the local community addressed in this curriculum?
4. Is there evidence of a consistent learning theory undergirding this curriculum? If so, what is the theory?
5. Does this curriculum have the potential to promote scientific literacy among children?
6. Are the instructional strategies proposed consistent with the ideals and goals presented in the curriculum?

Each of these questions needs to be addressed from two points of view: What is the answer to the question as claimed by the curriculum developers? What is the answer to the question as it is implemented in the design?

The process begins with construction of six *Analysis Forms* (a fancy name for worksheet), to correspond with the six questions. The analysis proceeds as each form is completed. The forms are numbered to correspond with the order in which they have been addressed in this chapter. There is no one correct way to proceed through the forms. Working on them one at a time requires examining the curricular materials six times, which becomes tedious, but does result in a very careful analysis of the materials. Trying to complete them all at once is complicated and results in a lot of paper shuffling. I found them most convenient to use two at a time, in this order. Forms 1 and 5 on the nature of science and scientific literacy complement each other; then come Forms 2 and 3 on the interests of students and community concerns; finally, we take up Forms 4 and 6 on learning theory and instruction. However, the forms will be described in the order in which the issues have been discussed.

To complete the forms, the person examining the curriculum needs to read carefully the introductory materials such as a cover letter and statements of purpose and goals to identify the rationale for the curriculum—what the developers claim will be included and why they have made the particular selection. Then the teacher needs to look at the implementation—what is actually included in the curriculum. It is not necessary to examine the entire curriculum guide. Two sections of the curriculum can be selected as samples of content. The curriculum developers may call these sections topics, units, or themes. The selection should include two sections that are different in some way. One may be a topic that has been taught recently and the other a topic that has never been taught. Or one may be a topic that was studied formally in college and the other one could be a subject not covered in college. Or the decision may be to select one topic from biology and one from physics. Or the first and last topics in the curriculum could be chosen. The two topics will be called the target topics.

Nature of Science. To analyze the curriculum for the nature of scientific knowledge that is presented, make a copy of Figure 1 to serve as Form 1. To check the claims of the developers, read the introductory materials that provide the rationale for the curriculum. When a statement is a claim about content, put a mark in the space labeled *Structural Component* on the diagram; when a statement is a claim about process and/or skill, put a mark in the space labeled *Procedural Component*; when a statement is a claim about the human implications or social aspects of science, put a mark in the space labeled *Human Component.* Statements that you perceive as being about more than one component (or statements that puzzle you) can be marked in a space that is the intersection of two or three circles. Your completed diagram might look like Figure 3.

Next, scan the two target topics for the stated objectives. They might be at the beginning of a lesson or in a sidebar on the page. Again, mark the copy of Figure 1 where you think each statement of objective might belong. It is helpful to make the marks different (change from x's to checks) or to use a

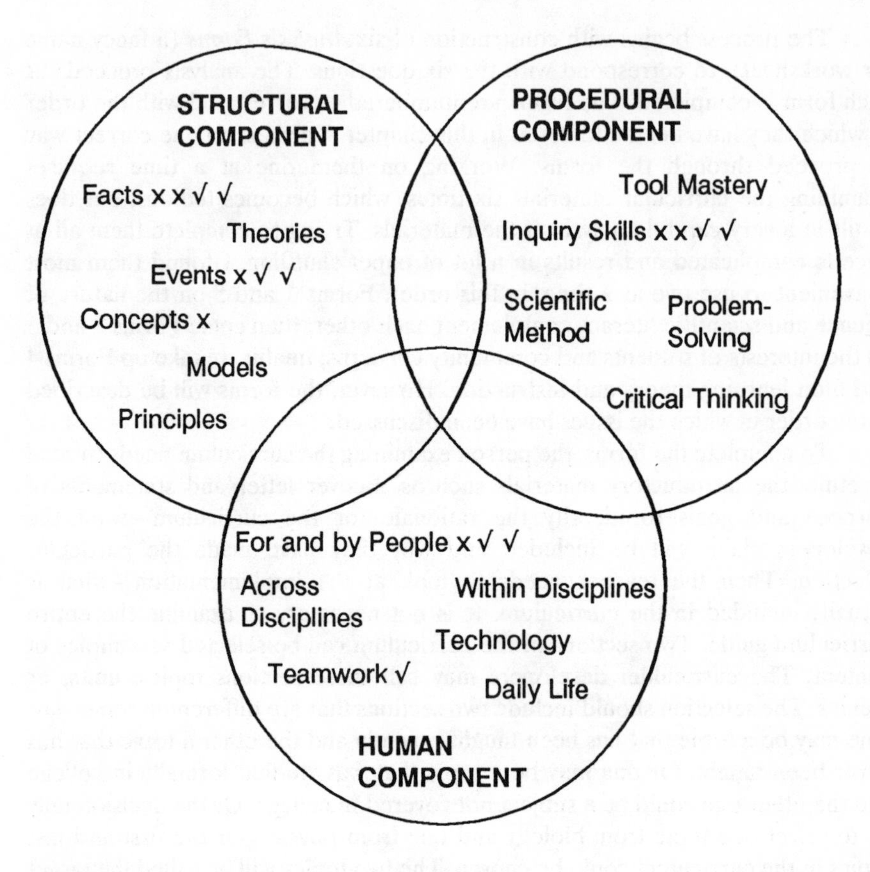

Figure 3
Analysis Form 1: Curricular Balance

different-colored writing instrument. There is no right answer in this examination, so do not spend a lot of time struggling with a decision about whether or not your mark is in the right space. It is safe to assume that your analysis of the nature of science from this diagram will not be isomorphic with an analysis completed by one of the curriculum's designers. (I would submit that the analysis diagrams of different developers from the same project would not be isomorphic either.) It is also safe to assume that the analysis diagram that you made is close to the picture of the nature of science that you are likely to teach from the claims of this curriculum. Do not be dismayed if you have no marks in the *bull's eye*. It is very difficult to incorporate all of the components of science into a single lesson at the elementary school level.

The last step in completing Form 1 requires review of the proposed instruction. While keeping in mind the students you usually teach, look at the two target topics you selected and read the directions for the instruction. In the appropriate places on the diagram, write phrases that describe what you would expect your students would learn from this lesson and what you imagine they might learn.

There is one more issue about a balanced approach to the nature of science (i.e., some structural knowledge, some procedural knowledge, and some human aspects of science) and the curricular design that needs to be addressed when examining the curriculum for its approach to the nature of science. Is the curriculum replete with events—activities in the typical language of a teacher? Are these events integral to the lesson and incorporated appropriately into the flow of the lesson(s) or are they tacked on at the end? A phrase or two about the events, their role, and placement at the bottom of Form 1 is usually sufficient to reveal the importance placed on events by the curriculum developers. These comments also can help to determine if there is a correspondence between the rationale stated by the developers and the curriculum as it is executed.

Problems, Interests, and Needs of Students. Examining a curriculum to determine if it addresses the problems, interests, and needs of students is not nearly as formal a procedure as the analysis of the structure of knowledge of science. The questions teachers ask themselves are based on common sense. To record the analysis, use a facsimile of Form 2, which is reproduced in Figure 4.

Are the topics, activities, and examples in the curriculum similar to the topics my students talk about on the playground? Are they similar to the events that happen in our community? Are the examples current without being caught in the trap of being trendy? Are the diagrams interesting and current? Is the teacher encouraged to adapt lessons for the specific needs of her or his students? Are there any recommendations for modifications of lessons for students with physical or learning disabilities? Is there anything—examples, language, activities, diagrams—that hints of sexism or racism? These questions need to be asked twice: once about the rationale that accompanies the lessons that make up the curriculum and once about the lessons that will be used by the teacher to implement the curriculum.

Community Concerns. To complete the analysis of the curriculum in terms of its attention to community concerns, make a facsimile of Form 3, shown in Figure 5.

Again, the questions that need to be asked to determine if the curriculum addresses the concerns of the community are not unusual or demanding. Are there opportunities and requirements for students to visit and participate in the work of the community, both individually and as a class? Are there suggestions for members of the community to visit the class? Are the problems posed in the lessons of concern to the community? Can the problems be addressed and/or answered in this setting? Are there activities that require and provide opportunities to do science at home? Are there suggestions about how to keep

Analysis Form: Problems, Interests, and Needs of Students	Rationale	Target Topics
Familiar to students?		
Likely in our community?		
Current, but not trendy?		
Diagrams and photos?		
Equal access?		
Sexism or racism?		
Physical/learning disabilities?		

Figure 4
Analysis Form 2: Problems, Interests, and Needs of Students

parents informed and involved in the activities of the science class? Are there opportunities available and/or is there encouragement to modify lessons to local circumstances? Are there directions to the teacher about topics that might be controversial, with suggestions about how to teach them? Again, the questions need to be asked twice, once about the rationale for the curriculum and then about the curriculum as a teacher might implement it.

Learning Theory. Form 4 provides a place to record information about learning theories. It might look like Figure 6.

The questions that are helpful in deciding if the curriculum is guided by a learning theory are simple. In the rationale for the curriculum, is a learning theory stated, explained, and exemplified? Is the instruction described by the curriculum's guide consistent with the learning theory? More telling, however, is an examination of the instructional aspects of the lessons to determine who is doing the work. If the nature of science and recent research in learning both indicate that scientific knowledge is constructed and learning involves constructing this knowledge, the instructional activity should put the burden on the student. The learning theory should be implemented as hands-on/minds-on science instruction.

Analysis Form: Community Concerns	Rationale	Target Topics
COMMUNITY Students visit and do work of the community individually and as a class? Community members visit the class? Problems that concern the community?		
HOME Science at home? Parents informed and involved?		
CONTROVERSY Modification possible? Controversial issues?		

Figure 5
Analysis Form 3: Community Concerns

Scientific Literacy. Form 5 provides a tool to analyze a curriculum to determine if its stated and actual goals are likely to promote literacy. This form, Figure 7, is a checklist derived from the minimal characteristics of scientific literacy described above.

Once again, read the rationale. If a statement claims that the curriculum will promote one of these abilities in a child, make a mark. Then examine the target topics for ways in which you can imagine your students attaining the abilities of a scientifically literate person as a result of their participation in the lessons described. Place a mark in the appropriate space.

Instruction. Form 6, Figure 8, provides the opportunity to record six aspects of instruction that should be considered when evaluating the descriptions for instruction that accompany the curriculum – models of instruction, unity and coherence in the topics presented for instruction, examples of a variety of instructional strategies, the match between curricular goals and instructional strategies, roles for the teacher, and techniques to implement the proposed instruction and monitor students' progress in learning.

Analysis Form: Learning Theory
Is a learning theory stated, explained, and exemplified? Describe it.
In what ways is the instruction consistent with the learning theory?
In what ways is the instruction NOT consistent with the learning theory?
Who is doing most of the work during instructional time (hands-on/minds-on)

Figure 6
Analysis Form 4: Learning Theory

Does the instructional strategy follow an instructional model? Cosgrove and Osborne (1985) describe several models for instruction in science education. In their view, the following "generative learning model" is most likely to result in students learning science. Their model begins with Phase One, *Preliminary,* in which the teacher ascertains students' views and students complete activities that identify their current views. In Phase Two, *Focus,* the teacher establishes the context and students become familiar with the materials that they will use to clarify concepts and describe events. In Phase Three, *Challenge,* the teacher facilitates exchanges of views and students test the validity of their own views. In Phase Four, *Application,* the teacher contrives problems that will allow students to test their views while the students attempt to solve the problems using their current views of the concepts. This is not the only model of instruction, but it suggests what a model of instruction might look like.

Second, is there some purpose to the topics and their sequence as they are presented for instruction? Third, does the instructional plan offer the teacher several options from which to select when deciding on the method of instruction for a lesson? Fourth, does the instructional strategy match the instructional goal and the instructional model? Fifth, is the role of the teacher defined clearly?

Analysis Form: Scientific Literacy	Rationale	Target Topics
Observe and describe		
Pose questions		
Use terms accurately		
Explain		
Concepts as constructs		
Process skills		
Predict		
Design experiments		
Work together		
Communicate		
Scientific knowledge as constructed		
By and for people		
Current events		
Daily life		
Enthusiasm		

Figure 7
Analysis Form 5: Scientific Literacy

Again, Osborne and Freyberg (1985) present a full description of the roles of a teacher in a classroom where scientific knowledge is being constructed. Lastly, are there suggestions for implementing the instruction and monitoring students' behavior and learning?

Analysis. Now lay out the completed forms and decide if the curriculum has the potential to enable students to become scientifically literate. There is no one correct resolution to this dilemma. Rather, it is a decision about what is most appropriate for these students in this community. Does the curriculum

Analysis Form: Instruction	
Describe the model of instruction presented.	Examples of instructional strategies that match the curricular goals.
Theme or themes that provide unity and coherence in the topics presented for instruction.	How to implement these strategies.
Examples of alternative instructional strategies.	Role of the teacher.

Figure 8
Analysis Form 6: Instruction

present a balanced picture of science as constructed knowledge? Does it address the problems, interests, and needs of the students? Is it possible to be responsive to the needs of the community through the use of this curriculum? Is the curriculum supported by a learning theory? Does it provide students with opportunities to experience science? Will the curriculum provide opportunities to develop the abilities of a scientifically literate person? Is the instruction congruent with the curriculum? How much adaptation will be required to make this nationally developed curriculum meet local needs?

Using the Process of Examining a Curriculum for Scientific Literacy: Three Examples

In discussing the new generation of elementary school science curricula, Bybee (1988) claims that they have four characteristics in common: (a) the content is organized on conceptual schemes such as Cause/Effect or Change and not by topics such as plants, (b) technology is integral to these curricula, both as tool and as content, (c) the philosophy that "less is more," offered by Morrison (1963), is the standard as the curricula attempt to cover fewer topics in greater depth and engender in students understanding, appreciation, and enthusiasm for science, and (d) the curricula are based on new assumptions about how children learn so that constructivism as a learning theory and cooperative groups as an instructional strategy are prominent.

Three Elementary School Science Curricula

At the time of this writing, the new curricula are in different stages of development. Therefore, I have selected three to use as examples of what might be found in analysis, using the process described above to determine the potential of an elementary school science curriculum to promote scientific literacy in children. They are: *Full Option Science System* (FOSS), being developed by the Lawrence Hall of Science in Berkeley, California and published by the Encyclopedia Britannica Educational Corporation; *Science for Life and Living (SLL),* being developed by Biological Sciences Curriculum Study, Colorado Springs, Colorado and published by Kendall/Hunt Publishing, Inc.; and *Improving Urban Elementary Science* (IUES), being developed by the Educational Development Center at Newton, Massachusetts and published by Sunburst Communications.[2]

FOSS consists of 16 modules for students in grades 3-6. It is supported by a *Materials Kit* from Delta Education, Inc., Nashua, NH. The materials that I received for analysis also included sources for books, video, and software as resource materials for both teachers and students. Each module includes sections on Overview, Background, Purpose, Materials, Cooperative Groups, Getting Ready, Students with Disabilities, Doing the Activity, Reflecting on the Activity, and Interdisciplinary Opportunities and Extensions. These same sections are repeated with specifics for each activity. I selected as target topics Learning Experience 1 in grade 4, "What Do We Already Know?" on surveying the uses of electricity, and Learning Experience 6, "Predictions #1," on wiring circuits.

SLL contains 24 modules plus introductory and integrative sections that cross modules and grade levels. They are designed for students in grades K-6. *Materials Kits* are provided by Science Kits, Inc., Tonawanda, New York. The *Teacher's Guide* contains an Introduction which includes an Explanation of Themes, the Curriculum Framework, the Instructional Framework, and the Lesson Format. The Lesson Format includes Title, Focus, Teacher's Role, Student Objectives, Estimated Time, Materials, Prepare Ahead, Before You Begin, Teaching Strategies, Check-up, and Resources. The organization of each module is displayed graphically. The contents of each activity in the module are presented as a concept map (Novak & Gowin, 1984) and the concept maps are developed as each activity in the module is presented. Forms and overhead transparencies are supplied. I chose as target topics the first activity in grade 5 on "Transformers," which uses pendulums, and the integrative activity on fast foods, "What's in the Bag?"

IUES consists of 17 modules for students in grades K-6. The materials I received for analysis included a list of resource books and places to contact for further information. The *Teacher's Guide* has four sections: Learning Experiences, Assessment, Teacher Resources, and Science Background and Glossary. The Learning Experiences section for each activity contains the following divisions: Overview, Learning Objectives, Suggested Time, Materials, Science Terms, Advanced Preparation, and Assessment. Staff development and supplementary materials for parents and school administrators are promised. I chose as target topics the first activity in grade 3, "The First Straw," on measurement, and the first activity in grade 4, "Color Writing," on chromatography.

Nature of Science. For FOSS, in the combined analysis of the rationale and two target topics, I made five marks in the structural component space, one in the procedural component space, and two in the human component space. For SLL, there were two marks in the structural component space, nine in the procedural space, and seven in the human component space. SLL also had one mark each in the oveslapping spaces and one in the "bull's eye." For IUES, there were three marks in the structural component space, six in the procedural component space, and three in the human component space. This analysis is not very informative in and of itself. All three curricula do address the three components of scientific knowledge, albeit in different ways. All of the instructional guides have students involved in activities—the events of science.

Problems, Interests, and Needs of Students. The FOSS curricular materials contain special notes for the instruction of students with physical and learning disabilities. Using rulers made from soda straws to introduce measurement seemed an especially effective way to help students understand that the process of measurement is a human construct. The integrative activity in the SLL curriculum on fast foods is surely a topic that is of interest to students. The use of comic strips and schematic drawings works well in catching the attention of students and is unlikely to become outdated quickly. The IUES introduction stresses the concern of the developers for low ability, minority, and special needs students living in urban areas. The module on "Circuits and Pathways" begins with uses of electricity and light bulbs, which is a common experience for all students. I must admit, however, that I have never seen anywhere a motor like the one used in the module, except in a science classroom and in some model kits.

Community Concerns. The IUES curriculum is particularly successful in addressing community concerns. It suggests visits to the classroom by a mechanic and by a spokesperson from the local utility company. Two of the activities require students to get information from their homes and their parents.

The recommendations for involvement in community concerns were not as clear in the other two curricula.

Learning Theory. Each of the three curricular guides examined acknowledge that children construct knowledge as the undergirding theory of learning. FOSS contains a 14-page document in the Teacher's Material on the "Biological Bases of Thinking and Learning." In the materials reviewed, all of the lessons began either with an event or a discussion of what the students know about the topic. All of them had students doing most of the work during the instructional time. However, all of the materials in the section on instruction indicate that the teacher is the person who poses the questions. This is of some concern because it is possible to infer from the guides that the right to ask questions belongs solely to the person in authority.

Scientific Literacy. All of the curriculum guides had marks in some rows and columns. None had marks in all the rows and columns. This is to be expected as different activities are intended to produce different learning outcomes. In my analysis of the materials available and used, SLL had the most consistency between the claims for developing the abilities of scientific literacy and the potential for their being achieved in the target topics.

Instruction. The FOSS curricular materials contain no explicit model of instruction. However, it is not difficult to infer a model from the document on "Learning Theory." The instructional strategy proposed is cooperative group learning, which is explained clearly. There is much support material for the teacher on how to structure and monitor cooperative groups. FOSS selects one module each in four areas – scientific reasoning and technology, physical science, earth science, and life science – as themes around which to organize the activities. The SLL curricular materials describe an instructional model of "Engage, Explore, Explain, Elaborate and Evaluate" and describe the role of the teacher in each stage. Every activity is matched carefully to the appropriate stage in the instructional model. Again, the instructional strategy of choice is cooperative small groups and much support material is provided. The modules are organized on the dual themes of "science, technology, and health" and "skill development." The IUES materials describe an instructional model of getting started, exploring and discovering, processing for meaning, and extending ideas. The relationship between the learning theory, the instructional model, and the role of the teacher is easy to see in the two clear figures that are provided. The instructional strategy is cooperative small groups; the guide explains clearly how to organize and monitor such groups. The IUES materials also contain suggestions for continuous assessment.

None of the guides offers alternative models of instruction besides cooperative small groups. On the one hand, I certainly have no intention of denigrating the use of this powerful instructional strategy. On the other hand, it is difficult for me to imagine any one strategy so powerful that it is the best for every student on every topic in every context. Memories of classroom experiences, both as a student and as a teacher, suggest that even the most stimulating and creative strategy becomes boring when overused.

From this report on the use of a process for assessing science curricula, it should be evident how much information can be gathered from a systematic exploration of a curriculum guide. In this cursory presentation of the analysis, it is also clear that the three sets of curricular materials reviewed have similarities and differences, advantages and disadvantages. None is perfect; all will require adaptations.

Caveat

Before bringing this paper to a close, there are some practical considerations that should be mentioned. One is how much it will cost to implement one of the curricula. Materials in kits can be purchased to supply the activities for all three of the curricula. Kits have obvious advantages and disadvantages. One advantage is that having the kit on hand reduces the amount of teacher preparation required. This may include the time needed to find the materials and cut, paste, or sort them into packets for students. The disadvantages include the cost of the kits, the fact that kits get damaged, pieces get misplaced or are used up, and administrators are reluctant to continue to put large sums of money into kits once the initial investment has been made. Also, a kit may be costly because it has materials for topics a teacher does not teach.

The biggest warning associated with the use of kits is the possibility that they might require too little exertion on the teacher's part. By having materials available without preparation, a teacher might adopt them without thinking (Apple & King, 1977). *Deskilling* is a danger in all of the curricula reviewed. They are all very directive, telling the teacher what to do and what to say. The SLL materials even indicate possible student responses. The danger of deskilling is a matter of no small consequence. First, deskilled teachers are perceived as not having the knowledge and skills to be decisionmakers for their own classes. Thus, deskilling does not promote professionalism. Second, when teachers know the answers to predetermined questions by consulting a curriculum guide, it should be expected that students and teachers alike will look on the printed word as the authority, not on events in nature as the reason for scientific knowledge. In such a case, scientific knowledge becomes a search for the right answers in the book, not the construction of personal knowledge. Using predetermined questions, always asked by the teacher, also deprives students of opportunities to learn how to frame questions.

It is not surprising that elementary school teachers, who are traditionally assigned the responsibilities for teaching reading, writing, spelling, arithmetic, social studies, art, music, and physical education are not experts in science. Their lack of expertise in teaching science is even less surprising when one realizes that many elementary school teachers have taken limited course work in science – and that was often in the form of introductory-level university courses in which science is taught as a rhetoric of conclusions, with a myriad of obscure facts to be memorized and as a subject that requires expensive and sophisticated equipment. There is a fine line between providing a teacher with the support needed to teach science well in order to promote scientific literacy and imposing a single "right way" to teach science. It requires the professional judgment of a teacher to recognize that a curriculum guide is just that, a guide, and to acknowledge its limitations. Recently, *Newsweek* magazine (Miller, 1989), in commenting on teachers' reactions to a packaged curriculum for teaching AIDS, said, "It is easy to use.... It's all packaged with materials that hold up and with lesson plans that are laid out the way teachers like." The same may be said of the elementary school science curricula that are being developed. Busy teachers with many responsibilities may have the knowledge and skills to transform and adapt science curricula to their students' needs, but may not have the time and opportunity to make such transformations. Scientific literacy is most likely to result from the integration of good curricular materials and good transformations made by good teachers.

Conclusions

This chapter on elementary school science curricula is deliberately naive. It has consciously avoided the controversial topics of textbook selection and assessment of student learning. Also, it is based on the assumption that teachers are recognized as professionals. Professionals have both theoretical and practical knowledge and skills. Professional teachers are concerned with the well-being of each student and with the class as a whole, know their subject matter and how to teach it, and reflect on and modify their practice. However, curricula development is not and should not be the prime responsibility of a schoolteacher. Teaching is sufficiently demanding in and of itself. Teachers need to rely on elementary school science curricula developed at the national level by committees with expertise in many areas. The responsibility of the professional teacher is to take these broadly-based curricula and modify, adapt, and transform them to meet the needs of individual students and classes of students. When these goals are attained, the potential of these curricula can be realized and students will have the opportunity to achieve scientific literacy.

Notes

1. The National Science Foundation-sponsored Publishers Initiative Program involves a partnership of a university (or scientists and science educators), a publisher, and a school district. Programs funded under this initiative thus far are: (1) *Science for Life and Living*, Biological Sciences Curriculum Study (Principal Investigator: Rodger W. Bybee); (2) *Improving Urban Elementary Science: A Collaborative Approach,* Education Development Centers, Inc. (Co-Principal Investigators: Karen Worth and Judith Sandler); (3) *Super*

Science: A Mass Media Program, Scholastic, Inc. (Principal Investigator: Victoria Chapman); (4) *The Life Lab Science Program: Development of a Comprehensive Experimental Elementary Science Curriculum,* Life Lab Science Program, Inc. (Co-Principal Investigators: Gary Appel and Roberta Jaffe); (5) *The Science Connection,* Houston Museum of Natural Science (Co-Principal Investigators: Carolyn Sumners and Terry Contant); (6) *National Geographic's Kids Network Project,* Technical Education Research Centers, Inc. (Principal Investigator: Robert F. Tinker); and (7) *Full Option Science System–FOSS,* Lawrence Hall of Science (Principal Investigator: Lawrence Lowery).

2. The decision to use these curricular materials for examples is based on the materials made available to the author and the ease in finding good examples to demonstrate the process. It is neither an endorsement of these materials nor a criticism of these curricula or the ones not mentioned.

References

Alexander, P., Bahret, M. J., Chaves, J., Courts, G., & D'Alessio, N. S. (1986). *Biology.* Atlanta, GA: Silver Burdett.

Apple, M. M. & King, N. R. (1977). What do schools teach? In R. H. Weller (Ed.), *Humanistic education* (pp. 27–63). Berkeley, CA: McCutchan.

Arons, A. B. (1983). Achieving wider scientific literacy. *Daedalus 112*(2), 91–122.

Biological Sciences Curriculum Study. (1988). *Science for life and living: Experimental Edition.* Dubuque, IA: Kendall/Hunt.

Bybee, R. W. (1988). Contemporary elementary school science: The evolution of teachers and teaching. In A. B. Champagne (Ed.), *This year in school science 1988: Science teaching; Making the system work* (pp. 153–172). Washington, DC: Association for the Advancement of Science.

Cosgrove, M. and Osborne, R. (1985). Lesson frameworks for changing children's ideas. In Osborne, R. & Freyberg, P. (Eds.), *Learning in science* (pp. 101–123). Auckland, New Zealand: Heinemann.

Dewey, J. (1934). The supreme intellectual obligation. *Science Education, 18*(1), 1–4.

Educational Development Center. (in press). *Improving urban elementary science: A collaborative approach.* Pleasantville, NY: Sunburst Communications, Inc.

Goodman, H. D., Emmel, T. C., Graham, L. E., Slowiczek, F. M., & Shechter, Y. (1986). *Biology.* Orlando, FL: Harcourt, Brace, Jovanovich, Inc.

Hirsch, E. D., Jr. (1987). *Cultural Literacy,* Boston: Houghton Mifflin Company.

Lawrence Hall of Science. (in press). *Full option science system—FOSS*. Chicago: Encyclopedia Britannica Educational Corporation.

Miller, M. (1989, June 5). Teaching kids to say no. *Newsweek*, p. 77.

Morrison, P. (1963). Less would mean more. *American Journal of Physics. 31*, 626.

National Board for Professional Teaching Standards. (1989, January 13). *What teachers should know and be able to do* (Draft). Oakland, CA: Author.

Raizen, S. A. (1987). Assessing the quality of the science curriculum. In Champagne, A. B. & Hornig, L. E. (Eds.), *This year in school science 1986: The science curriculum* (pp. 179–206). Washington, DC: American Association for the Advancement of Science.

Raizen, S. A., & Jones, L. V. (Eds.). (1985). *Indicators of precollege education in science and mathematics.* Committee on Indicators of Precollege Science and Mathematics Education, National Research Council. Washington, DC: National Academy Press.

Shamos, M. (1988). (1988, November 23). The flawed rationale of calls for "literacy." *Education Week*, p. 28, 22.

7

Scientific Literacy: Perspectives of School Administrators, Teachers, Students, and Scientists From an Urban Mid-Atlantic Community

Nancy W. Brickhouse, Diane Ebert-May, and Betty A. Wier[1]

Everyone seems to agree that Americans should be scientifically literate and that the promotion of public understanding of science is a "good thing" (Shortland, 1988). This, however, is where consensus begins and ends. Definitions of scientific literacy are ubiquitous and elusive, not only among educators but also among politicians, business people, and other citizens. Although there is no consensus on what it means to be scientifically literate, schools are accused of producing graduates who are scientifically illiterate. The data that point to our *illiteracy* in science consistently imply that science education in the schools is failing (Koshland, 1989; National Research Council, 1989; Schneps, 1988; Triangle Coalition, 1988; and Bybee, 1985).

If schools are to develop scientifically literate graduates, it is necessary to achieve consensus on a definition of scientific literacy first. Although the definition is likely to be multifaceted, it would make explicit the expectations for the results of teaching science in the schools and, would provide a basis for the development of state and local curricular guidelines, textbooks, tests, in-service education of teachers, and pre-service educational programs of undergraduates. These factors would enable schools to meet the public's expectations for a scientifically literate high school graduate.

The goal of the discussions reported here is to contribute toward efforts to define scientific literacy. To that end, four roundtable discussions were held on this topic in the spring of 1989. Individuals representing three different professions in our urban, mid-atlantic community as well as university students participated; all of whom have been active in science education in the schools. Many are recognized by their peers as leaders in science education. All indicated keen interest in the project. They were: (a) administrators from a local school district, (b) scientists (retired or practicing industrial chemists), (c) university students (undergraduates), and (d) K-12 teachers from local schools.

The school administrators' group (individuals responsible for coordinating a school district's K-12 curricula) included an administrative assistant to the superintendent of schools, an elementary school principal, a supervisor for elementary school instruction, a supervisor for secondary school education, and a biology teacher who is the chairman of the science department at his high school.

The teachers' group included six elementary school teachers, one junior high school science teacher, and two high school chemistry teachers. We chose to include a larger number of elementary school teachers in this roundtable group because local attention is focused on science education in the primary and intermediate grades.

Five of the six undergraduate students who participated in the discussions are elementary education majors; however, only one has completed her science methods course. All have completed their science requirement of twelve credit hours. The sixth student is a history/biology major in the college of Arts and Sciences.

Most of the scientists have retired from a local industry. Three of them are visiting science teachers in elementary schools, one supervises graduate students at the local university, and one is currently employed as a research chemist in industry.

The discussions were moderated by a staff member of the American Association for the Advancement of Science who posed the question for discussion: "What does it mean to be scientifically literate?" and probed into the statements of the participants. Both video- and audio-recordings were produced at each discussion. Then the tapes were transcribed and the responses were coded according to the categories described below. Some of the transcripts were coded independently by two or three of the authors with an interrater-reliability of 0.87. The responses for each group were analyzed to determine the frequency with which individuals mentioned aspects of each of the categories.

Aspects of scientific literacy were categorized according to a plan that was adapted from Roberts' (1983) system to categorize science teaching. Another category of scientific literacy, Appreciation of Science, frequently mentioned by participants in the discussions, was added to accommodate aspects that did not fit into any of those proposed by Roberts. Roberts' system can be used to divide scientific literacy into seven categories of analysis:

1. *Everyday Coping* comprises practical applications of scientific knowledge and reasoning skills that enable effective functioning in society. These skills enable application of science to the management of personal needs with regard to health, nutrition, and safety or the use of mechanical devices found in the home.

2. *Structure of Science* includes knowledge about the nature of science and how it functions as an intellectual discipline. Understanding the nature of science involves the ability to examine the relationship between evidence and theory, the adequacy of models to explain phenomena, and how scientists develop new knowledge.

3. *Science, Technology, and Decisions* implies a working knowledge of the features that distinguish science and society, and technology and society as well as their interrelationships. This knowledge and its related skills are applied to make decisions and solve problems where science, technology, and society interface.

4. *Scientific Skills* encompasses processes of scientific inquiry—observing, classifying, measuring, making hypotheses, experimenting, drawing conclusions, problem-solving, and higher order thinking skills.

5. *Correct Explanations* infers an understanding of the ends of scientific inquiry—facts, principles, and correct explanations of scientific phenomena—which have value for their own sake, not because they are useful immediately.

6. *Self as Explainer* category refers to the cultural context of scientific explanations. It includes an understanding of how scientific ideas were developed and applied as a function of human purpose as well as how scientific ideas are applied in conjunction with other types of information such as values and beliefs drawn from ethics, religion, and even superstition.

7. The *Solid Foundation* classification is related to the utility of scientific information as the substrate for further study of science. It recognizes that literacy implies the capacity to learn more science which, in turn, depends on having an understanding of basic scientific concepts.

8. *Appreciation of Science* (the additional category) covers characteristics of scientifically literate people—specifically, a keen interest in and a positive attitude toward science.

Statements made in response to the question, "What does it mean to be scientifically literate?" were categorized using this system. Then the frequency of statements in each category was compared across the groups.

Findings

Our analysis revealed important differences between and similarities among definitions of scientific literacy expressed in each group. The information in Table 1 summarizes the findings. The eight components of scientific literacy described above are listed and followed by an example of a statement that was assigned to that category. Categories are ordered in the table (and in the ensuing

Table 1
Ranking* of Categories of Scientific Literacy by Administrators, Teachers, Students, and Scientists

Category Sample remark	Group Administrators	Teachers	Students	Scientists
I *Science, Technology, and Decisions* "...there are environmental decisions that have to be made, nuclear decisions that have to be made...."	High	High	High	High
II *Scientific Skills* "...I want my students to understand and be able to perform the scientific processes of observation, making hypotheses, testing hypotheses, drawing conclusions, and making inferences."	High	High	Low	High
III *Everyday Coping* "ability to function in the modern world."	High	High	High	Low
IV *Correct Explanations* "You have to know the basic laws of science and you have to know some facts about the subject you're going to talk about before you can come anywhere near doing most of the other things."	Low	High	High	Low
V *Appreciation of Science* "...the excitement and the enthusiasm for knowledge that these children have which is certainly one part of what you want to have in scientific literacy."	None	Low	High	High

Table 1 (continued)

Category Sample remark	Group Administrators	 Teachers	 Students	 Scientists
VI *Structure of Science* "We [scientists] try to come up with ideas that tie everything together, but they're always inadequate. Any kind of scientific theory is not a completely accurate description of the world. We refine things as we get more evidence and more sophisticated theories."	Low	None	Low	High
VII *Solid Foundation* "Learning should be a progressive thing where knowledge is built on knowledge from one grade, where one grade has to prepare you for the next grade in science, and so on...."	Low	None	Low	Low
VIII *Self-As-Explainer* "...uses and processes of science as opposed to religion and magic."	Low	Low	None	None

* The terms "High," "Low," and "None" represent the relative importance of the category indicated by each group. The relative importance was gauged by observing the percentage of individuals in each group who made statements pertaining to that category.
High: No. $\geq$ 50%; Low: 0 < No. < 50%; None: No. = 0.

text in this chapter) according to the frequency with which they were mentioned. The frequency ratings were determined by calculating the percentage of individuals in each group who made statements pertaining to that category. They are reported as high – mentioned by at least 50% of the discussants; low – mentioned by less than 50% of the discussants; none – not mentioned by any of the discussants.

Table 1 indicates that the three categories mentioned most frequently to describe scientific literacy were: Science, Technology, and Decisions; Scientific Skills; and Everyday Coping. The two categories mentioned least frequently were Solid Foundation and Self-as-Explainer. With regard to group similarities and differences, there is more congruence between the responses made by the teachers and the administrators than between any other groups. The greatest disparity is observed between the scientists and all of the other groups.

We will discuss the most frequently mentioned categories of scientific literacy first and then address those referred to less frequently, noting important rationales, values, and assumptions used to support statements about the importance of the aspect discussed.

Science, Technology, and Decisions

Individuals in all four groups commented frequently on understanding science-related social issues as an important aspect of scientific literacy. All groups assumed that science education focused on societal issues would produce a more informed voter and a more reasoned decisionmaker.

In spite of the agreement on the importance of understanding science-related social issues, there was disagreement on why this was so critical. The administrators, students, and teachers stated repeatedly that the primary reason this aspect of scientific literacy is important is because of the harm being done to the environment.

> Student 1: I think you have to [understand science] to understand the workings of the world and the environment in order for us to progress as a country without destroying ourselves. I think that it's very important that everybody is scientifically literate so that they know what goes on when they turn on the water and let it run while they're washing their dishes instead of turning it off.

> Teacher 1: But my question is, is it more important for a kid to know why it's hot in the summer and cold in the winter, or what to do about it? In other words, it's cold in the winter, so I'm going to need to do certain things to stay warm. If I do these things correctly, number one, I'm going to save the environment; number two, I'm going to save myself some money; number three, I hope I'm going to create a better environment, a better living situation.

> Administrator 1: [A]s I'm walking toward the ocean, I say, "Oh gracious. I need to understand what's going on in the ocean, the implications for how we're abusing our waters...."

The environmental concerns expressed by the teachers, students, and administrators were those that had been reported recently in the local newspaper with such headlines as: "Chemical Industry Wrote the Laws We Live By," "What a Mess: We're Wasting the World with a Plague of Plastic Debris," "OSHA: Its Business is Shielding Business," and "When Earth Becomes a Hothouse." So it is not surprising that these groups were aware of and concerned about environmental issues.

Compared to the students, administrators, and teachers whose remarks were clearly influenced by the popular press, the scientists' reasons for emphasizing science, technology and decisions as a component of scientific literacy were that literate individuals would be less influenced by what they read in the newspapers so that they would make informed decisions when voting.

> Scientist 1: One of the reasons that it's important that everybody have some reasonable technical background and knowledge of scientific principles and information [is] so that they can make intelligent decisions other than what they read in the newspapers on important factors that come before us in this modern world. Without that information and knowledge, the voters are going to follow whatever blind precepts they can pick up from other informed sources such as the press, oftentimes, and are going to make decisions on an unscientific basis.

None of the scientists expressed concern about harming the environment. One chemist's primary concern was about society's influence on science.

> Scientist 1: [S]ociety probably has more influence on science than science does on sociology, on society....I think the voter concepts [are] so much more important; that is, the misconception, for example, that some of our methods of trash disposal are harmful. And they may not be, or vice versa.

The transcripts reveal that the scientists differ from the other professional people with regard to their perspectives on the political and economic ramifications of environmental and safety issues. (The role of economics in science was mentioned by only one other individual, a teacher.)

> Scientist 2: I think of the demonstrations against nuclear power plants like in New England, $5 billion sitting out there and they're hung up on the fact that there's not a safe evacuation route.

> Scientist 1: Everybody's got the notion now that nuclear energy is bad. And yet, when you analyze the facts, why you may not come to the same conclusion. If the facts are presented clearly, people may come to the conclusion that nuclear energy is good by comparison with all the known alternatives.

A third scientist proposed that the most appropriate way to approach the topic of nuclear energy is by making comparisons between the amount of pollution generated by coal and oil versus atomic energy.

The comments in these transcripts suggest that an important facet of scientific literacy in the view of some of the scientists is support for science-related industry—the chemical industry in the case of our community. The other groups either did not acknowledge the role of the chemical industry in our local educational system or they did not believe that industry was primarily concerned with its own agenda.

> Administrator 1: I get the feeling by their industry doing that [developing new products that pose fewer environmental threats], which is my interpretation, that they are saying to our youngsters who are in school, yes, we did what we felt we knew based on the scientific information available at that point, but now we're finding out that what we did was not the best thing for the environment and we as chemists are going to remedy the situation. So I don't get the feeling that they would want us to do anything different from what we think we should be doing as morally and educationally correct in teaching youngsters that maybe not everything they hear, even though it may come from a science-oriented company, is necessarily true and that perhaps they should investigate further.

Many agree that being able to make decisions about socioscientific problems is essential to being scientifically literate. However, the reasons for these emphases are very different for the scientists than for the other three groups. These differences could impose a substantial barrier to achieving consensus on how scientifically literate individuals think and act. For example, is the public's decision not to allow a nuclear power plant to become operative a result of scientific literacy or scientific illiteracy? Teachers, administrators, and students believe that scientifically literate individuals should be concerned about safety issues relating to nuclear energy because they have an understanding of radioactivity. If scientists believe that scientifically literate individuals should be less concerned about safety issues than about economic and environmental trade-offs, then we may have irreconcilable differences in conceptions of how scientifically literate people should behave relative to the domain of public policy decisionmaking.

Scientific Skills

A component of scientific literacy mentioned frequently by the administrators, teachers, and scientists and occasionally by the students, was the development of scientific skills. A reason for valuing this aspect of scientific literacy is that these skills can be transferred to solving problems outside the classroom.

> Administrator 1: [W]hen we get into the scientific area, I think we're looking at something very important, and that's the scientific process, and what that brings to individuals which is transferable to all areas of their life.

> Teacher 2: I think what you want people to do is to question things and not only be able to ask the questions, but then know how to answer them or know whether the information they're getting makes sense. If they've had experiences doing those kinds of things, they might be better equipped.

In addition to teaching thinking skills and hoping for transfer to new situations, the scientists, who perceive science as a process, thought that the development of scientific skills should be taught because "that is what science is all about" and students should experience the process of struggling and testing their own ideas.

> Scientist 3: I think the important thing is that you have to teach science as a process, as a way of acquiring knowledge – acquiring information and converting it into knowledge. That's what science is all about. It's trying to understand the world around us.

> Scientist 4: One thing which I think is very important and that is, often, science is taught with the idea that the student is supposed to get the answer and everything is headed toward trying to find the answer. And you say, what is the teacher really looking for here? What I am trying to give the teacher, to give in turn to the students, is that they should explore and struggle and come up with their own formulations of how things work and test their own ideas.

Not surprisingly, the scientists' rationales for science instruction were based on their experiences as practicing scientists. The rationales of the other groups arose more often from their thinking about people than about science. Once again, this reflects the nature of their work, which is centered on working with people and on what information they think people need to function in daily life.

Although the teachers asserted that scientific skills were an important aim for science teaching and that we should promote the transfer of scientific skills to everyday problems, they also voiced some skepticism about how easily this can be achieved.

> Teacher 3: [P]erhaps somebody who does something intuitively comes up with a solution for that problem, but might not be able to

transfer it to another situation because it isn't seen as part of a process....

Teacher 2: In order to perform those processes, you've got to have a certain body of knowledge and a certain set of skills anyway.

Like the scientists whose daily experiences in the laboratory determine their views on the nature of science, teachers' daily experiences in the classroom provide them with reasons to believe that teaching general skills is not quite as simple as it may appear to others.

The students' failure to emphasize the development of scientific skills is an interesting finding because such development is a valued goal of science education. This failure can be explained in part because, generally, the students based their ideas of scientific literacy on their own experiences as students and, usually, learning scientific skills is embedded in a lesson to demonstrate a particular scientific idea. In our experience, teachers are more likely to describe the intent of a laboratory activity by the concept they want students to understand rather than by the intellectual and process skills that will be performed and increased. Therefore, students may be unaware that developing these skills is an instructional goal.

There is considerable agreement that the development of scientific skills is a significant part of scientific literacy and no one expressed any opposition to emphasizing its importance in instruction. The students, administrators, and teachers believed it was important because of the role of scientific skills in helping people to solve everyday problems. The scientists also thought it was important but for a different reason: because it is an important component of science. Regardless of the different explanations given for the need to acquire scientific skills, the perspectives are not contradictory because the goals toward which they are aiming are the same. The difficulty educators may have with this component of scientific literacy is in knowing when it has been achieved; that is, knowing when scientific skills are being used to solve everyday problems, because science as currently taught makes very little attempt to help students transfer scientific skills to real-life situations.

Everyday Coping

The ability to apply scientific knowledge to everyday situations was mentioned frequently by the students, administrators, and teachers, but less frequently by the scientists. One of the teachers gave an example of everyday coping as a description of the behavior of a person he considers to be scientifically literate.

Teacher 3: I was changing a tire for a friend. I was on the ice, so I knew that this was a situation involving action and reaction. In other words, if I pushed one way, my legs might go the opposite way and I'd be on the ground. But my friend was oblivious to this concept.

This category was stressed particularly by the students. Their reasons for this emphasis were based largely on their own experiences and their feelings that this was an important issue that was missing from their science courses. They stressed that people should understand scientific phenomena that have bearings on their daily lives and that teachers are responsible for making the connection between the information taught in science classes and real-life situations, rather than leaving students to make all the connections on their own.

> Student 2: [M]y situation was that I learned the elements and I learned different chemicals, their names and characteristics. But I never could understand why somebody didn't take packages from the store and say; "Look at this; you know, these [chemicals] are in here."

The students also submitted that understanding human health and weather was important because these are topics they found interesting and relevant in their science courses. When the students were asked what knowledge about the weather really mattered to people's daily functioning because the weather cannot be controlled, they were unable to give an answer.

The administrators and students stressed the importance of being able to use high technology more than the other groups. Their reason was that it might help students to function in society and in future jobs.

> Student 1: If you learn how to work a computer so you can go out to McDonald's and get a job, then great, that has taught you something, and that's scientific literacy.

Although the teachers emphasized the importance of being able to function in the modern world, they made no specific recommendations of what kind of knowledge is necessary to do this. (One teacher suggested that we are actually functioning rather well, considering we are all supposedly scientifically illiterate.) Two of the scientists also mentioned repeatedly the importance of making science relevant to people's lives.

> Scientist 1: You point it [science] toward practical things that people can identify with.
>
> Scientist 3: That's right; into their own lifestyle, their own problems.

Their specific examples of content that would help in this area were ordinary acids, bases, and polymers—the specialty area of one of the scientists.

Once again, there is considerable agreement in all groups that scientific literacy should focus on those aspects of science that people are likely to need in their daily lives. The difficulty with reaching consensus in this area is in answering the question: What specific content do people need to learn in order to function in today's world? The administrators and students provide a clue in their suggestions that technology and health are important.

Correct Explanations

Despite the fact that all groups placed considerable emphasis on learning the science necessary for everyday coping, learning the traditional content of science for its own sake was cited frequently by teachers and students as important and occasionally by scientists and administrators.

> Teacher 1: You have to know the basic laws of science and you have to know some facts about the subject you're going to talk about before you can come anywhere near doing most of the other things.

It is noteworthy that this aspect was emphasized by the two groups who have spent the most time in science classrooms in the last several years. Traditional science teaching emphasizes this aspect. Students and teachers see value in the tradition and believe that people should understand certain scientific principles even though they may never use them. It is reasonable to expect that many of these teachers, who could be described as traditional science teachers, have focused a great deal of their instruction on correct explanations. Therefore, there may be a degree of self-esteem at stake when they assert that the content they have been teaching for the last several years is important.

This issue did raise some controversy among the teachers, however. Most concluded that certain scientific principles, such as the reasons for seasons, plant growth, the solar system, and the nature of energy, should be known by a scientifically literate individual, although no one offered explanations for why these particular ideas were recommended over others. One teacher suggested that she wanted students to master the objectives on the state and local school district curriculum guidelines. However, this drew dissent from another teacher.

> Teacher 2: Sometimes, in a curriculum, it's very difficult to include some of the higher-level thinking skills that are so necessary for science. What has happened is that we identify the things that are most easily measured and that drives our curriculum so that if we can test children and say they can tell what the four seasons are or [what] a specific body of knowledge is then we feel we've done an adequate job. And I think that's where we're missing the boat.

In the administrators' discussion of which concepts should be taught, there was concern that the local school district's guidelines were often ignored and that the content of courses was determined by the textbook.

> Administrator 2: I would say that what determines the curriculum more than anything else is the textbook that the teacher selects, not the state curriculum and not the district curriculum, but the textbook.

Two of the scientists stated that having a scientific vocabulary is important to scientific literacy and that this vocabulary is a prerequisite to understanding science.

> Scientist 3: You have to have a basic vocabulary of scientific and technical terms and concepts.

> Scientist 1: If you can speak the language to some extent, then you have the feeling that you can understand it.

Another disagreed, however, and recalled having been transferred to a different job and being told that he was now a textile chemist. He joked that he then had to learn how to spell "textile."

Teachers and scientists also expressed concern about the number of errors found in newspaper articles dealing with science and their desire for people to be able to recognize such errors.

> Scientist 1: I think that's what happens in the newspapers today; there's so much erroneous material appearing, partly innocently because the people who are called upon to do the writing aren't equipped to handle the subject competently. So they write what they think or what they believe or what they don't know and people read it and say that [because it] appears in the newspapers it must be true.

> Scientist 4: I very seldom see an article on science in a newspaper that doesn't have at least one error in it.

> Teacher 4: Just in the past couple of weeks, I became upset with an article in the local newspaper, simply [because of] what I think was incorrect usage of terms.

The teacher then referred to an incident in which the press reported that when a cargo door came off an airplane, the people were "sucked out" and another report that claimed that 20 pounds of calcium had been applied to slick streets to melt the ice. The scientists related an incident in which the press had used the word "melt" to describe a dissolving process. Both groups were questioned about why they thought these were such important errors. A teacher gave an "everyday coping" rationale; he stated that he wanted his students to recognize that calcium should not be put on public roads because it does not improve the condition of the roads. However, neither group could justify why the other two errors made a difference to the meaning of the news stories because the net effects were the same, regardless of the terminology used.

Once again, there is reasonable consensus that scientifically literate individuals should know some of the correct explanations of science. However, there was no consensus on which explanations are so important that everyone should know them, but it can be inferred that explanations that impinge on daily life are more important than those that do not. The correct explanations that were mentioned by the various groups probably would not qualify as being major ideas in science (except for one teacher's suggestion of the nature of energy). The students could not name a concept that tied their science courses together. They described the purpose of the science courses as being "just to give you a real surface understanding." In the absence of consensus on which correct explanations are important for scientific literacy, it is reasonable to expect that they will continue to be determined by textbook publishers who create the materials used in most science classrooms.

Appreciation of Science

All groups recognized that if students enjoyed science they would be more likely to want to learn about it. However, only the scientists and the students suggested frequently that this was an important part of scientific literacy. They saw appreciation of science not as a means of obtaining scientific literacy, but as an integral part of it. A few of the elementary school teachers also said that scientific literacy should include appreciation of science, but none of the administrators suggested it.

One reason the students gave for believing that appreciation of science is important is that, if they enjoyed science, people would be more likely to participate in it after they leave the classroom. The scientists were more concerned that people be curious because "if you're curious, you'll seek the right answers; if you're not curious, you'll accept whatever you see." Both of these rationales suggest that the goals of scientific literacy include learning that extends beyond the classroom.

> Scientist 2: So our glory is that a lot of us keep our curiosity all our lives. If we encourage people to be curious and show them new things and so on, we can increase the number who remain curious.

In discussing the need for people to like science more, the scientists often reflected on how they became interested in science and how they would like others to share their enthusiasm.

Many may speculate that scientists would also desire a citizenry that loves science so that it will be more supportive of investments in scientific research. Although this rationale has been expressed elsewhere (Shamos, 1988), there is no evidence that this was a concern of this particular group of scientists.

All of the teachers and students who emphasized appreciation of science as a part of scientific literacy were either presently elementary school teachers or were training to be elementary school teachers. The secondary school teachers rejected this idea. It is notable that this emphasis appears in elementary schools, where more attention is given to the affective needs of students, whereas in secondary schools the emphasis is often entirely on the cognitive needs of students.

Although the scientists gave much consideration to this aspect of scientific literacy, it is apparently not important to the other groups. Most considered appreciation of science useful only because of its power to motivate students to learn science in school, not for its own sake.

Structure of Science

The scientists were the only group that stated frequently that there were important aspects about the nature of science that need to be included in the education of the scientifically literate individual. Teaching the nature of science was brought up in the administrator and student roundtable discussions also, by individuals working toward or having degrees in scientific disciplines.

Aspects of the nature of science that the scientists believed are crucial for everyone to understand included the dynamic nature of science, the purpose of searching for patterns in nature, and the scientific attitude and method. The idea that there are patterns in nature that may be discovered was deliberated by several of the scientists. One scientist explained that by repeatedly observing this orderliness, one comes to realize that "there is a certain bunch of rules that all of nature follows." However, a third scientist expressed the view that people need to understand that science is constantly changing and is not a list of rules to be memorized.

> Scientist 4: Often, in people's first look at science, it looks like some monumental thing.... These are the rules. These are the laws. This is what you have to pound into your head – without getting the idea that science is a dynamic thing that's constantly changing – and our views of the world are changing as we get more information. We try to come up with ideas that tie everything together but they're always inadequate. Any kind of scientific theory is not a completely accurate description of the world. We refine things as we get more evidence and more sophisticated theories.

One scientist advocated understanding "the processes and methods of science for testing models of reality" and "the scientific attitude" as the most important concepts for all students to know. His view of the scientific method includes: conducting experiments, recording observations, organizing information, determining similarities and differences in information, and trying to reach conclusions. He defines the scientific attitude as having a strict regard for accuracy, controlling experiments, being honest, and being impartial.

In contrast to the empiricist view of the scientist, one administrator, also a biology teacher, advocated a more human-centered understanding of science.

He remarked that there are characteristics of science, such as tentativeness and uncertainty, that he wants his students to understand. His reason is that understanding that science cannot always produce a single right answer is particularly important in dealing with crucial science-related social issues.

One student mentioned that people should comprehend the tentative nature of science so that they will not be surprised when they learn that scientists have changed their minds on a particular conclusion. If people understood that changing one's mind, based on new evidence, was a part of the scientific endeavor, they might have a more accommodating attitude toward scientists.

In these discussions, the only individuals who felt that understanding the nature of science should be a part of scientific literacy were those with strong backgrounds in science. Although the scientists did not explain their reasons for its importance, the teachers and students believe it is important so that people can interpret better the complex societal problems and current reports of scientific and science-related social issues in newspapers. Again, the rationale is premised on everyday coping.

Reaching consensus on what should be taught about the nature of science could be quite difficult for two reasons. One, the scientists frequently mentioned this aspect in their discussions, but it was rarely if ever mentioned in the other deliberations. Two, although the scientists believe that there are aspects about the nature of science that they want everyone to know, they do not agree on the aspects and some of their views are very different from what current philosophers of science would argue are the characteristics of science.

The influence of logical positivism is seen in one scientist's views on teaching science to a child. He argues that one can begin with the idea of the scientific method and the scientific attitude, but then "get into the carefully researched accurate explanations of why things are not the way they seem, but *this is the way they really are.*" This is very different from the Kuhnian view (Kuhn, 1970) expressed by another scientist that scientific theories are always inadequate and require refinement.

Gaining consensus on the nature of science is exacerbated because many of the ideas expressed about the scientific method and the scientific attitude are incongruent with contemporary analyses of the nature of science (Brown, 1977; Feyerabend, 1975; Kuhn, 1970; Lakatos, 1970; Toulmin, 1972). Although a thorough investigation of this issue is beyond the scope of this paper, we examine one point in particular. One scientist said: "Scientists are always honest. They always record exactly what they see, not what they think they should see." First, there is considerable evidence that scientists are not *always* honest and occasionally report data never observed (Lewin, 1989; Marshall, 1989). Second, what we see is dependent on what we think we ought to see (von Foerster, 1984). This is a limitation of science, not an issue of honesty. Because we cannot make an infinite number of observations or process every stimulus we sense, we choose to make certain observations and attend to particular details based on our theories of that which is most important.

If we cannot agree on the *nature of science* or on whether or not it is important for the scientifically literate person to understand the structure of science, then we have essentially no common ground from which to work. General agreement on this issue will require protracted debate.

Solid Foundation

Providing students with a foundation of knowledge on which to build in other courses was not an important issue in any of the discussions. Most participants limited their comments to individuals who were not likely to take science courses in college; therefore, being prepared for subsequent science instruction was moot. The way in which science instruction needs to be structured in order to provide continuity between grades prior to college is a curricular matter beyond the scope of the roundtable discussions. However, individual scientists and students referred to the need for a logical progression of learning from grade to grade. The importance of the knowledge base developed in the K-12 years as the foundation for continuing to learn science outside the formal academic setting was not raised in the discussions.

Does the fact that preparing high school graduates for lifelong learning in science was not discussed mean that the requisite knowledge base is not an aspect of scientific literacy for the high school graduate? Even if it is, the question about the nature of the foundation necessary to be prepared for the vicissitudes of post-high-school life is left unanswered. Clearly, the public debates raging over scientific illiteracy indicate that a solid foundation has not been provided.

Self as Explainer

Understanding science in its cultural context and cognizance of how nonscientific beliefs and understandings – ethical and religious values, for instance – relate to science were not emphasized to a great degree in any of the groups. They were addressed by one administrator/biology teacher and two teachers as having relevance to their instruction.

The biology teacher/administrator suggested that bioethical decisionmaking and other value-laden issues should be part of science education. Although no one opposed this idea, the other participants immediately raised the concern that incorporation of these topics could be perceived by community members as indoctrination. The other participants were concerned about how a population ignorant of science can affect the scientific enterprise adversely. For example, some participants in the administrators group were concerned that people may decide to forbid certain debatable topics based on irrational fear rather than for informed, logical reasons. At the same time, administrators were interested in gaining public support for their district's work.

> Administrator 1: From the committee in our district dealing with religious concerns, I get a very strong feeling that they it would want us to teach both sides of an issue very, very carefully and make certain youngsters know that there are positives and negatives to whichever alternative they choose.

One teacher expressed the position that science must be distinguished from religion and magic, but did not detail why he thought this was important. Another teacher asserted repeatedly that we must examine values, especially when discussing such issues as the environment. He maintained that his students now have a great deal of knowledge about garbage. However, he is skeptical that this knowledge will transfer into action without addressing values explicitly in the classroom.

Because understanding science in a societal context was mentioned by only three individuals, all teachers, it seems that most of the discussants do not consider this a substantive aspect of scientific literacy. We suspect it will continue to be taught in their three classrooms solely because these teachers believe it is very important, but widespread teaching of this aspect will have to wait until it is incorporated into curricular guidelines and instructional materials.

Summary

We were able to establish a number of areas in which there was considerable consensus on what we ought to be teaching to create a scientifically literate citizenry. The groups generally agreed that it is important to focus more on science-related social issues, on scientific knowledge that people need in order to manage their daily lives, and on problem-solving skills. Unfortunately, these are rather tenuous agreements when examined in light of the disagreements on the same issues. For example, how will we know when we have a scientifically literate population that is capable of making decisions about science-related social issues? What knowledge of science do people really need to function in today's world? What scientific problem-solving skills are most useful in everyday life?

There are also ways in which the voices of the groups are unique, as would be expected given the differences in their training and their daily experiences in science or education. The students' comments were based on their experiences as learners in science classrooms. Their major criticism of their science courses was that they are irrelevant to their daily lives. They had more difficulty than the other groups in forming coherent arguments about scientific literacy, as might be expected in view of their status as novices.

The teachers expressed more disagreement with one another and more skepticism for easy answers than the other groups. They, as well as the students, also see more value in traditional science instruction than the other groups and, therefore, may be more cautious about change than the other groups.

The scientists often based their arguments of what should be taught on what they believe science is and how they became interested in the subject. However, the administrators were primarily concerned about the students' abilities to function in the community and about accountability in science programs.

Our study focused narrowly on groups in a single community who are actively involved in science education in various ways. The groups are not intended to be representative of all students, teachers, administrators, or scientists. Some of the issues raised probably reflect the local culture. Few communities have a heavier dependence on the chemical industry than the one in which these individuals live. The community also has several, very active environmental groups. In addition, some local newspaper reporters appear to have been effective in raising the consciousness of the community about environmental problems. Much of the discussion about scientific literacy focused on the environmental concerns of the community and on the need to understand better the functioning of the local industries. Because they live in the same community they share many interests and concerns. Yet, the different groups' reasons still show fundamental differences that pose substantial hurdles to reaching unanimity on scientific literacy.

Given these difficulties, reaching national consensus on scientific literacy will be challenging. At the national level, we need to bring together individuals from all parts of the country, who are likely to have diverse local problems, to determine the role local issues play on the different emphases of scientific literacy. Additional groups such as parents, academic scientists, and businesspeople need to be included in the discussions in order to get a clearer view of what these groups mean when they use the term "scientific literacy."

We believe we should not wait for national consensus before beginning to work toward attaining scientific literacy at the local level. Promoting dialogue with various groups provided a basis for identifying the concerns of the constituents in our community and may be a starting point for achieving a degree of consensus. The resulting actions to improve scientific literacy, although multifaceted in nature, should begin to meet the community's expectations that the schools will produce more scientifically literate high school graduates.

Notes

1. The authors are listed in alphabetical order. We wish to thank Betty J. Calinger, Audrey B. Champagne, and Barbara E. Lovitts of AAAS for moderating and assisting in the roundtable discussions.

References

Brown, H. I. (1977). *Perception, theory and commitment: The new philosophy of science.* Chicago: Precedent Publishing, Inc.

Bybee, R. W. (1985). The Sisyphean question of science education: What should the scientifically and technologically literate person know, value, and do – as a citizen? *NSTA Yearbook. Science Technology Society.* 79–93.

Feyerabend, P. (1975). *Against method.* London: Berso.

Koshland, D. (1989). Science competency through fun. *Science, 24*(243), 989.

Kuhn, T. S. (1970). *The structure of scientific revolutions* (2nd ed., enlarged). Chicago: University of Chicago Press.

Lakatos, I. (1970). Falsification and the methodology of scientific research programmes. In I. Lakatos & A. Musgrave (Eds.). *Criticism and the Growth of Knowledge.* Cambridge: Cambridge University Press.

Lewin, R. (1989, April 21). The case of the "misplaced" fossils. *Science, 244*(4902), 277–279.

Marshall, E. (1989, June 23). UCGS reports a fraud. *Science, 244*(4911), 1436.

National Research Council. (1989). *Everybody counts: A report to the nation on the future of mathematics education.* Washington, DC: National Academy Press.

Schneps, M. H., producer (1988). *A private universe.* [Videotape] Santa Monica, CA: Pyramid Film & Video.

Shamos, M. H. (1988, Oct. 3). Science literacy is futile; try science appreciation. *The Scientist,* p. 9.

Shortland, M. (1988). Advocating science: Literacy and public understanding. *Impact of Science on Society, 152,* 305–316.

Toulmin, S. E. (1972). *Human understanding.* Princeton, NJ: Princeton University Press

Triangle Coalition for Science and Technology Education. (1988). *The Present Opportunity in Education.* Pittsburgh: Author.

von Foerster, H. (1984). On constructing a reality. In P. Watzalick (Ed.). *The Invented Reality* (pp. 41–61). New York: W. W. Norton.